Energy & Security
in the
Industrializing
World

Energy & Security
in the
Industrializing World

Edited by
RAJU G.C. THOMAS
and
BENNETT RAMBERG

THE UNIVERSITY PRESS OF KENTUCKY

Written under the auspices of the
Center for International and Strategic Affairs,
University of California, Los Angeles.
(A list of other Center publications appears at the back of this book.)

Editorial and Sales Offices: Lexington, Kentucky 40508-4008

Library of Congress Cataloging-in-Publication Data

Energy & security in the industrializing world / edited by Raju G.C.
   Thomas and Bennett Ramberg.
      p.   cm.
   Includes bibliographical references.
   Includes index.
   ISBN 0-8131-1699-6 (alk. paper)
      1. Energy policy—Developing countries—Case studies.   2. Energy
industries—Political aspects—Developing countries—Case studies.
3. Developing countries—National security—Case studies.
4. Developing countries—Foreign relations—Case studies.
5. Nuclear weapons—Developing countries—Case studies.
6. Developing countries—Military—Case studies.   I. Thomas, Raju
G. C.   II. Ramberg, Bennett.   III. Title: Energy and security in the
industrializing world.
HD9502.D442E49   1990
333.79′091724—dc20   90-12513

Contents

Figures and Tables

1

The Relationships among Energy, Security, and the Economy

RAJU G.C. THOMAS

With the oversupply of oil on world markets and the fall in international oil prices from $32-41 per barrel in 1980 to mid-1989 levels of $15-20 per barrel, the international energy crisis that followed the 1973 Arab-Israeli war appears to have passed.[1] The world has shifted from devising strategies of crisis management to strategies—albeit rapidly fading—for avoiding similar crises in the future. Such were the lessons learned from the sustained energy crisis of the 1970s, which severely affected both the industrialized and developing countries for almost a decade.

For the Western industrialized nations, economic dependency and vulnerability in the hands of a small number of developing countries, mainly in the Islamic Middle East, came as a rude shock. In the crisis, only a radical reevaluation of Western diplomatic and strategic policy in the Middle East could keep the oil flowing to the West.[2] Attitudes toward the Arab-Israeli dispute, particularly the Palestinian issue, had to be adjusted in the West—especially in the United States—so as not to alienate the conservative Arab oil-producing states. Whereas few Americans had previously known the difference between Persians and Arabs, or between Shiite and Sunni Muslims, these distinctions were quickly learned. Monarchies such as Iran and Saudi Arabia assumed considerable importance, both as major suppliers of petroleum and as markets for massive sales of military hardware, intended to reverse the flow of petrodollars.

The competition among the industrialized weapons suppliers, especially between Eastern and Western bloc countries, and the sudden accumulation of advanced weapons among oil-exporting countries of the Middle East threatened to upset the military balances that had prevailed among Israel, the conservative and radical Arab states, and Iran. At the same time, the prolonged oil crisis established the economic and strategic interdependence of the Western industrialized countries and the Islamic

Middle East. It became clear that conflicts and domestic political up-heavals in the Middle East could not be ignored, since every major disturbance implied the threat of a disruption in the oil flow to the West.

The economic consequences for the developing countries were no less severe, although the Western nations had substantially more prosperity at stake. For example, oil demand in the United States in 1973 was 28.61 barrels of oil equivalent per capita, compared with only 1.38 barrels per capita for the Less Developed Countries.[3] Nevertheless, in many indus-trializing countries, such as India, Pakistan, and Brazil, economic shocks from the oil crisis led to severe foreign exchange shortages and the curtail-ment of various development programs. Unlike wealthy countries, low- and middle-income states were unable or unwilling at the time to trade arms for oil to correct their trade imbalances. (The notable exception is Brazil, which more recently has managed to step up its overseas sales of small arms and ammunition.) Instead, some of these states resorted to other economic tactics. For example, by encouraging unskilled and semi-skilled labor to work in the Middle East oil-producing countries, India, Pakistan, South Korea, and Taiwan were partly able to offset the high cost of oil imports through petrodollar remittances from their "export" labor forces. Despite the economic near-catastrophes suffered as a result of the much higher oil prices demanded by the oil-exporting countries of the Middle East, the diplomatic strategy of the developing countries took on an unexpectedly supportive role for the Organization of Petroleum Export-ing Countries (OPEC), in the hope of obtaining special economic con-cessions and trade and investment benefits.

An offshoot of the 1970s energy crisis was the belief by some of the developing countries that a long-term solution could be found by embark-ing on or accelerating nuclear energy programs. The dramatic drop in oil prices that began in the mid-1980s has not necessarily reversed the com-mitment to nuclear energy development in many industrializing states. The choice of this energy alternative, especially by countries with prevail-ing or perennial security fears—among them India, Pakistan, South Af-rica, South Korea, and Taiwan—has renewed the concern that resources and capabilities acquired in the nuclear energy sector could make more difficult the control of nuclear weapons proliferation.

While the world in 1989 is no longer in the midst of an energy crisis, it should be clear that there are underlying relationships among (a) a nation's energy needs and external dependency; (b) its economic and political stability; and (c) its broader security concerns. The intensity of these relationships will, of course, vary from country to country in the developed and developing worlds, and within each country over time. Perhaps in the current framework of an international oil oversupply and low prices, the basic relationships may appear obscure, and "crisis prevention" inappropriate. Much of the new mood of optimism arises from oil discoveries, increases in energy efficiency, especially through advances in computerized automation, and the prospect of tapping alternative energy sources. However, none of these conditions appears to guarantee the long-term resolution of the energy problem. Lessons and legacies forgotten are no less relevant. For example, the rapid consumption of oil discovered in the Gulf of Mexico, the North Sea, and Alaska since 1974 only underlines the inevitable limitations of world oil reserves. Energy conservation through automation also has its limits. While computer technology may fine-tune energy usage in vehicles and buildings, such systems, costly in themselves, can only slow, not halt, the escalating demand for energy. Nor is there yet any clear alternative energy with the exception of atomic power, which in itself constitutes part of the "energy-economy-security" problem.

When dealing with security in the energy context, we are concerned with a broad and an unavoidably subjective connotation of the term. Such a maximalist interpretation encompasses economic, political, strategic, and military security as against the more familiar minimalist interpretation that focuses on military threats and defense programs alone. Economic security thus suggests national resource sufficiency and, in particular, access to goods and services in world markets at affordable terms. Political security suggests the maintenance of domestic stability, whether based on rule by the consent of the governed or on various degrees of authoritarian measures; either way, law and order prevail and economic, political, and social activities are conducted with little or no hindrance. Strategic and military security is partly outward-looking and may be guaged by the degree and intensity of perceived external threats and the military capabilities that can be marshaled to meet those threats. It is also partly inward-looking, in that it involves the diversion of domestic resources and services to meet those threats.

The energy crisis of the 1970s struck at the heart of all three forms of security concerns. For many industrializing countries, including India, Brazil, Pakistan, and South Korea, the periodic spurts in OPEC oil prices produced severe imbalances in foreign trade and severe dislocations in domestic economic plans. Whereas at one time the oil import bill was about 20-30 percent of their export earnings, it quickly rose to some 50-80 percent by the mid-1970s.[4] Oil shortages and higher prices hit key economic sectors, including the fertilizer and plastics industries and air and road transportation services. The resulting inflation aggravated these countries' economic troubles.

Economic crises invariably lead to political instability. As the people of a nation feel the effects of economic stress, they are likely to vent their anger on the government in power, whether democratic or authoritarian. The Indian democracy collapsed with the declaration of the "emergency" between 1975 and 1977 because of economic stresses and public dissatisfaction. On the other hand, authoritarian regimes in South Korea, Pakistan, and Brazil began to feel the pressures of political liberalization and potential revolution as a result of the economic crisis at home.

Strategic and military concerns arise more indirectly as oil exporters accumulate arms either to protect their economic resources and new-found prosperity, or simply because they can afford to buy sophisticated arms with surplus petrodollars, whether needed or not. Security issues arising from the new arms race are complicated by the competition among arms sellers from both the Western and Eastern bloc countries, which results in client military states such as American-backed Saudi Arabia and Iran under the Shah on the one hand, and Soviet-backed Iraq and Libya on the other. Oil importers, especially the advanced industrialized states, may also choose to deploy their military capabilities to protect their international energy sources and supply lines.

Instability may arise not only from new strategic relationships between the regional energy exporters and the internationally powerful energy importers but also from the relationships between developing countries themselves. Many of the new security concerns have been confined to the Middle East and South Asia, for which the oil crisis produced strategic interdependence. The massive purchase of arms from both West and East by oil-exporting states such as Iran, Saudi Arabia, Iraq, and Libya also contributed to the Indo-Pakistani arms race. Concerned about Pakistan's military links with Saudi Arabia and with the Shah's Iran, and apprehensive about the general arms buildup in the Middle East and possible arms transfers to Pakistan, India speeded up its decisions on weapons acquisitions from the industrialized states and other decisions on general force modernizations at home.[5] The Indian focus on nuclear energy programs also raised the prospect of diversions to a nuclear weapons program and of Pakistani counterresponses to such Indian intentions.

Energy crises, in short, strike at both the economic and defense sectors. Energy policy options among the industrializing countries must, in turn, be adapted or even overhauled to take into account the dual effects on security and the economy, because the balance struck between defense and development programs is tied to fluctuations in the energy market, both at home and abroad. Defense and development programs invariably call for expensive imports of machinery and machine tools, and foreign exchange for such imports is depleted by undue dependence on foreign energy supplies.

The energy factor thus plays a crucial role in the traditional "defense versus development" debate in the industrializing world. This debate usually revolves around the impact of defense expenditures on the national economy or, conversely, how economic conditions affect a nation's ability to embark on costly defense programs. There are two broad perspectives on the defense-development debate.[6] On the one side are those analysts who depict the opportunity costs of defense programs in terms of the development plans of Third World states. On the other side are those analysts who emphasize the spinoffs of defense spending that are favorable for the economy through economies of scale in civilian sectors, mainly the electronics, aerospace, automotive, and shipbuilding sectors. However, the relationship between defense and development as affected by international and domestic energy pressures is seen in the constraints on defense purchases abroad because of depleted foreign exchange reserves, and the constraints on defense procurement at home in a sluggish and unstable economy caused by energy shortages.

In sum, it should be clear that a nation's energy policy and management carry significant implications for both its security and economic domains. Energy shortages at home require adept diplomacy and adequate bargaining power abroad to fill the breaches. External and internal security, as well as external trade policies and economic development plans, have roots in the successful or unsuccessful management of energy policy. While the international energy crisis may appear to lie behind us, the basic linkages among energy, security, and the economy remain, and energy policy management must aim at either maintaining the present equilibrium or advancing to safer levels. As fossil fuels are rapidly depleted, energy crises will take new forms, and the current search for energy alternatives at home has tended to point to nuclear energy as the potential solution, which in itself poses security problems for the next decade.

Figure 1.1 delineates the manner in which issues of energy, security, and the economy affect the broader aspects of defense and development planning. Such an analytical approach begins with an assessment of available domestic and external energy resources. Indeed, the degree of a country's energy surplus, self-sufficiency, or deficiency is the determinant of the other issues of this study. On the supply side, a nation must determine its actual and potential domestic energy resources and then measure them against its actual and potential demand for energy. The shortfall between supply and demand will indicate the degree to which a country must rely on overseas purchases of energy.

External energy dependence raises problems of both accessibility and affordability. The question of accessibility concerns political obstacles that may have to be faced by a country in dealing with energy-exporting

Fig. 1.1.   Energy impacts on security and the economy

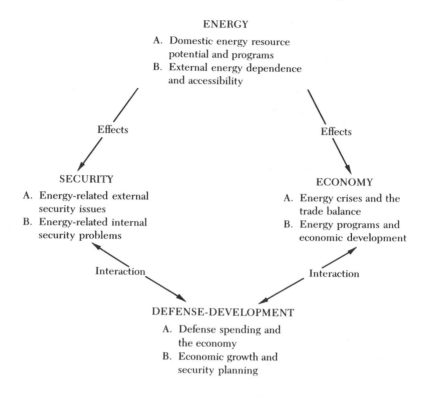

ENERGY

A. Domestic energy resource
   potential and programs
B. External energy dependence
   and accessibility

Effects   Effects

SECURITY   ECONOMY

A. Energy-related external   A. Energy crises and the
   security issues   trade balance
B. Energy-related internal   B. Energy programs and
   security problems   economic development

Interaction   Interaction

DEFENSE-DEVELOPMENT

A. Defense spending and
   the economy
B. Economic growth and
   security planning

countries. Two significant cases are Israel and South Africa. Indeed, the beginning of the Arab oil embargo to the West and the ensuing worldwide oil crisis was provoked by the 1973 Arab-Israeli war and Western policies on the Palestinian issue. Neither Israel nor South Africa has had access to Arab oil at any time in the past, and the fall of the Shah of Iran in 1979 terminated Iranian supplies of oil to both countries. Politically more isolated, South Africa had to rely on purchases from "underground" suppliers that bypassed the international economic sanctions supported by almost all Third World countries, and on alternative domestic sources of energy. India, too, faced some difficulty. Despite its consistent support for the cause of the Palestinians, India had to engage in greater diplomatic maneuvering among the OPEC members of the Middle East than its Islamic neighbor, Pakistan, to ensure the steady supply of oil and oil products. Diplomacy became especially critical after the 1981 outbreak of war between Iran and Iraq, the two countries of the Middle East that had traditionally supplied much of India's oil imports. Shortfalls had to be supplemented with extensive energy imports from the Soviet Union.

Similar problems were faced by countries such as Brazil, Argentina, Taiwan, and South Korea, all of which have had to cope with varying degrees of external energy accessibility and affordability. However, supplies from OPEC countries in areas of lower tension than the Middle East—Venezuela, Ecuador, Nigeria, Gambia, and Indonesia—as well as the emergence of Mexico as a major exporter of oil outside the OPEC bloc, helped to temper the political and economic severities of the oil crisis during the crucial 1974-83 decade. In the case of Cuba, which has been more isolated than most other Latin American countries, economic dependence on the Soviet Union was further increased. One of the lessons of the oil crisis era is that political accessibility to overseas sources of energy cannot always be taken for granted.

Economic affordability depends on the importing country's ability to negotiate favorable prices, or—where such prices are officially non-negotiable because of the OPEC cartel's price fixing policies—to negotiate favorable lending and supply terms and to obtain other economic relief. The economic negotiating skills of nearly all of the energy-importing countries are taxed to the fullest in a crisis.

The effects of energy pressures are thus felt both in terms of security and in the economic arena. As delineated in figure 1.1, problems of external security arise from arms races fueled by petrodollars, or from temptations to use military force to resolve some of the repercussions of acute energy shortages that may cripple a country's economy. Threats to internal security arise from domestic violence or revolutionary movements caused to some degree by energy-related economic dislocations and depressions. In the economic arena, governments must monitor the impact of severe trade imbalances and foreign exchange shortages on development programs if they are to minimize the adverse economic consequences.

The broader relationships among energy, security, and the economy may be considered from two dual, interacting perspectives by each country attempting to resolve the problems arising from these relationships (fig. 1.2). From the first perspective, each state must make two assessments, one relatively objective, the other more subjective: (1) the extent to which it is dependent on external energy resources, both short- and long-term; and (2) the intensity of external threats that it faces. The level of external dependence on energy resources, and the severity of threats to its national security, may be perceived to range on a scale from high to low.

The second perspective concerns the policy management of energy and security whereby each state must attempt to achieve two goals: (1) minimize the adverse impact of external energy dependence on its economy; and (2) minimize the threats it perceives to its national security. Such policy efforts may be perceived as having proved either a success or a

Fig. 1.2.   Basic relationships among energy, security, and the economy

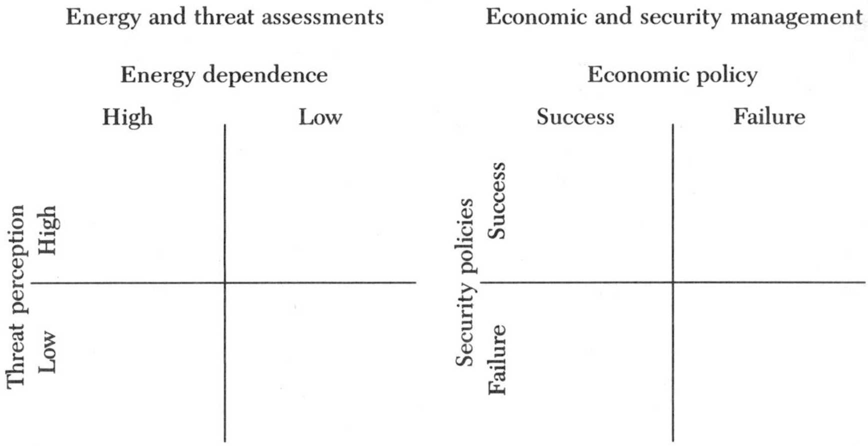

Energy and threat assessments   Economic and security management

Energy dependence   Economic policy

High   Low   Success   Failure

failure, again with variations in between. Needless to say, both the threats perceived and the degree of energy dependence, as well as the nature of national economic and security management, are much more complex than this two-part diagrammatic analysis. For simplicity's sake, however, the two levels of analysis may be conceptualized under the two matrices in figure 1.2.

Within the loose frameworks suggested here, the degree of external energy dependence may be seen as relatively higher in the cases of South Korea, Taiwan, Pakistan, Cuba, and Argentina than in the cases of India, South Africa, and Brazil, which have achieved modest levels of energy self-sufficiency. Of these countries, Pakistan and South Africa may perceive greater external threats to their national survival than do Brazil, Argentina, or even India. The successes or failures of policy responses under such different conditions are also varied and mixed. Despite high-level threats and high external energy dependence at the commencement of the oil crisis, South Africa has perhaps more successfully managed these problems through economic and security policies than Pakistan, although Pakistan's case can by no means be considered a failure. We are dealing here with degrees and intensities among different states, the assessment of which must necessarily be largely subjective. The two matrices essentially serve as conceptual reference points whereby states within the energy-economy-security triangle may be addressed.

The theme of this book, then, is the effects of domestic energy shortages and external energy dependencies on the security and economic policies of certain key industrializing states in Asia, Africa, and Latin America.

Issues confronting these states are analyzed in the dual context of crisis and post-crisis. Eight such states have been selected for this study: South Korea, Taiwan, India, Pakistan, South Africa, Cuba, Brazil, and Argentina. They represent varying levels of economic development but generally fall under the two broad categories of middle-income and low-income countries. All of these states were severely dependent on oil imports at the height of the oil crisis in the 1970s and have since embarked upon or stepped up nuclear energy programs at home.

The countries included in this study may be grouped under three categories.[7] The first group consists of India, South Korea, and Taiwan. At the height of the international energy crisis, the economies of these states were hit hard by soaring oil import bills. All three states have also embarked on nuclear energy programs.[8] Given the concurrent security concerns in their regions, this development has heightened international fears of nuclear proliferation. There are both similarities and differences in the domestic energy situations of the three nations, in their economic conditions at home, in their regional security concerns. Whereas all three were hard-hit economically at the height of the international energy crisis, only India had the domestic resource potential to overcome its dependency on external energy sources. By contrast, much of the coal in the Korean peninsula is found in the north rather than in the south, and Taiwan has few energy resources of its own. On the other hand, vigorous external trade policies on the part of South Korea and Taiwan, which have vastly improved their trade surpluses, have combined with the current abundance of oil in the international market to stabilize the "energy-economy" relationships in these two countries. These positive economic developments, however, have not slowed down their nuclear energy programs.

The security fears and the temptations to acquire a nuclear weapons capability in South Korea, Taiwan, and India arise from threats perceived in mainland China's growing nuclear weapons capability, although more immediate security fears in South Korea and India arise from North Korea and Pakistan, their neighbors; no doubt perception of a Chinese threat is much less than in earlier decades. Note also that South Korea and Taiwan have signed the 1968 Nuclear Non-Proliferation Treaty, while India has not, but signing the NPT may not provide sufficient clues about future policies, since article 10 of the treaty allows any signatory state to withdraw by giving three months' notice. The more pertinent questions are, What are the potential nuclear weapons capabilities of these states? and How strong are the security motives in these countries that may compel them to acquire nuclear weapons in the future?

South Africa and Pakistan fall into a different category than the first three, although they share certain similarities. It might be argued that there are far greater internal and external security pressures in South Africa and Pakistan and consequently greater temptations to acquire nuclear weapons. South Africa's abundant coal resources are no guarantee of energy availability at all times, since the working of coal mines is dependent on a hostile black labor force, and the nation has no major petroleum resources. On the other hand, nuclear energy appears to be a viable energy source, given South Africa's abundant uranium resources—the second largest in the world after those of the Soviet Union—and its self-reliance in uranium enrichment technology since the mid-1979s. South Africa is of particular concern for this study, given its minority white population surrounded by a hostile majority of blacks, its minority status as white-ruled state surrounded by hostile black states, and its confrontation by a hostile Third World that includes Arab members of OPEC. The linkages between energy and security are prominent in South Africa.

The significant question is what defense or deterrent purpose a South African nuclear weapons program would serve even if the country could divert its nuclear energy program in that direction. Unlike Pakistan, which faces the potential of a nuclear India, and unlike Israel, which conceivably faces the acquisition of nuclear weapons by states such as Syria, Iraq, Saudi Arabia, Egypt, and Libya, the white regime of South Africa need fear no such threat from the black African states—with the possible exception of distant Nigeria. A South African move to acquire nuclear weapons would serve largely to reinforce domestic psychological confidence, although such possession would also carry credibility as a threat against surrounding black states that cannot retaliate in kind.

The Pakistani situation is different. Although Pakistan has substantial natural gas resources, its coal and oil resource potential is limited, and its hydroelectric development potential is restricted by the high siltation levels of the Indus River. Pakistan's close diplomatic relations with Saudi Arabia and the Persian Gulf kingdoms—together with income received from the stationing of troops and military advisers in Saudi Arabia and the "export" of a large Pakistani labor force to the Gulf—have enabled Pakistan to stabilize and invigorate its own economy for the time being. However, Pakistan's search for energy self-sufficiency as manifested in its nuclear energy programs has raised serious international doubts about its ultimate objectives. Unlike the nuclear programs of India, South Korea, or Taiwan, Pakistan's was perceived to emphasize acquisitions, first of reprocessing technology and then of enrichment technology, that would provide it with a weapons capability. These capabilities—though legitimate objectives for a nuclear energy program—were sought before a serious nuclear energy program was implemented, raising questions about Pakistan's ultimate intentions.

In the energy context, a third set of countries consists of Argentina and Brazil, which resemble the others in many ways both during and after the international energy crisis. These two countries were hard-pressed to obtain petroleum from OPEC, given their rapidly declining foreign exchange reserves, but were steadily able to overcome such pressures by seeking alternative sources of oil supplies and by exploiting domestic resources. While coal resources in Argentina are insignificant, Brazil has meaningful quantities, although hardly sufficient to meet its total energy needs. Like India, however, both Argentina and Brazil have modest petroleum resources, the development and production of which have grown substantially during the last ten years. Argentina also has significant quantities of natural gas, which has begun to contribute to its energy self-sufficiency.

The difference between these two countries and those in the other two groups is the absence of major security concerns in Latin America. The unexpected Falkland Islands war between Argentina and Great Britain in 1982 may yet provoke Argentina into a nuclear weapons program, given the nuclear weapons of its adversary in that war. That in turn would surely raise security fears in Brazil that would compel it to counter the Argentinian program. For the present, the likelihood of another Anglo-Argentinian war appears remote, and conditions have returned to those that prevailed before the 1982 war. However, the major nuclear energy programs of both Argentina and Brazil raise concerns about possible diversion to weapons capabilities, if for no other reason than that of national pride and international prestige. Strangely, despite the absence of immediate security fears between Brazil and Argentina, the escalation of nuclear energy programs in both states has followed an action-reaction spiral that is characteristic of arms races, suggesting the possibility of a latent nuclear weapons race between the two. Although both Brazil and Argentina have signed the Treaty of Tlatelolco, which seeks to establish a nuclear weapons–free zone in Latin America, Argentina has not ratified it, and Brazil has not incorporated the treaty as part of its domestic law, leaving subsequent governments an opening to ignore its provisions.

Cuba does not fit any of the above three categories, but its significance in this study arises from a number of factors: the security concerns that Cuba's actions raise in the western hemisphere, the major Soviet economic and military involvement there, the Cuban decision to embark on a nuclear energy program, and its refusal to sign the Treaty of Tlatelolco. Cuba has only modest domestic oil resources; much of its liquid energy must be imported, nearly all of it from the Soviet Union. Cuba has no coal resources, and its hydroelectric potential is limited by the size of its rivers. For the present, the Cuban energy program being set up with Soviet

RAJU G.C. THOMAS

12

nuclear power plants is not of concern, but the transfer of Soviet technical knowhow may make possible the diversion to weapons production in the future. Moreover, the safety standards of the nuclear power plants to be supplied by the Soviet Union are not yet known. If Chernobyl-type accidents occurred at Cuban nuclear power plants, they could threaten the physical safety of the United States along its southern rim, as well as parts of Mexico, Central American, and the Carribean islands. Especially noteworthy is that Cuba has not signed the Treaty of Tlatelolco or the Non-Proliferation Treaty.

This book examines the crisis and postcrisis energy conditions and policy approaches of eight key states in the industrializing world. The focus is dual: energy impacts on each state's national economy and on its national security. The parallels and variations discovered in the experience of the eight selected states may well serve as models for national policymakers in these and other countries.

NOTES

1. See *New York Times*, Nov. 18, 1988.
2. For two studies at the height of the energy crisis, see Joseph S. Nye and David A. Deese, eds., *Energy and Security* (Boston, Mass.: Ballinger Publishers, 1980); and International Institute for Strategic Studies (London), *Energy and Security*, ed. Gregory Treverton (Westmead, England: Gower Publishing, 1980).
3. Stuart Sinclair, *The World Petroleum Industry: The Market for Petroleum and Petroleum Products in the 1980s* (London: Euromonitor Publications, 1984), p. 35.
4. Assessed from World Bank, *World Tables 1987* (Washington, D.C.: World Bank, 1987).
5. See Raju G.C. Thomas, "Aircraft for the Indian Air Force: The Context and Implications of the Jaguar Decision," *Orbis* 24, no. 1 (April 1980): pp. 85-102.
6. For a comprehensive study of this problem, see Nicole Ball, *Security and Economy in the Third World* (Princeton: Princeton University Press, 1988).
7. These categories were first developed in Raju G.C. Thomas, "India's Nuclear and Space Programs: Defense or Development?" *World Politics* 38, no. 2 (January 1986): pp. 315-19.
8. See James Everett Katz and Onkar S. Marwah, eds., *Nuclear Power in Developing Countries* (Lexington, Mass.: Lexington Books, 1982).

2

India

RAJU G.C. THOMAS

The influence of energy issues on India's security and development policies first took on significance in 1974 following two important events, one
external and the other internal. Externally, the quadrupling of oil prices by
the Organization of Petroleum Exporting Countries toward the end of
1973, from about $3 per barrel to $12 per barrel, threw into disarray the
commencement of India's Fifth Five Year Plan in 1974 and compelled
India's economic planners to revise and stretch the Fifth Plan over a period
of six years. Internally, India's test of an underground atomic device in May
1974 brought into question the ultimate purpose of its nuclear energy
program, which has, before and since, been declared as serving peaceful
domestic purposes only.

Today, more than 15 years later, the conditions that India faces are
fundamentally different. Instead of depending heavily on oil imports,
India now produces more than two-thirds of its oil needs. Its economic
performance has improved dramatically since international oil prices began to drop from highs of $41 per barrel in 1980 to mid-1989 levels of $15-20.
India is now the dominant military power in the South Asian region, with
arms capabilities that rival those of the oil-rich states. The transformation
in India's strategic setting has changed the nature of the issues confronting
India and the emphasis and direction of its policy-making. While the basic
components of the energy-economy-security equation remain, their relative importance and the nature of their relationships have changed.

The two major events of 1974—oil pricing shocks and the Indian atomic
test—highlight some of the linkages of energy, security, and development
policies in India. Heavy dependence on oil imports from the Persian Gulf
states, Libya, and the Soviet Union in the 1970s had implications for Indian
diplomatic, economic, and military policies. Apart from the obvious need
not to alienate these states if oil was to keep flowing on affordable commercial terms, it was important for India to step up exports to the oil-supplying
countries to prevent a foreign exchange crisis that would have crippled the
Indian development program. There was also a military dimension to the
international oil crisis. The Middle East oil-exporting states were some of

the largest buyers of sophisticated arms from the industrialized countries. The proximity of some of these states to the subcontinent, and the overt or latent military links of states such as Saudi Arabia and Libya with Pakistan, seemed at the time to have strategic relevance in assessing the security of India. Special purchases of oil and military equipment from the Soviet Union also accentuated India's economic and military reliance on Moscow.

While the international oil crisis of the last decade eroded India's security environment and adversely affected its development programs, India's nuclear energy program, in turn, has raised fears in Pakistan of an imminent nuclear weapons threat in the region. In the West, moreover, the atomic test of 1974 sparked a general fear of the collapse of the state of affairs that had been tentatively promoted with the passage of the Nuclear Non-Proliferation Treaty in 1968. Even though India and several other prominent countries still have not signed the treaty, its very existence was expected to exert pressure on such states not to embark on a nuclear weapons program. While the international energy crisis affected India's strategic environment and its economic development programs, India's nuclear energy policy in turn has affected the regional and global security environments.

The passing of the international energy crisis by the mid-1980s has brought greater economic prosperity in India, both because of successful efforts to tap oil resources, especially in the offshore Bombay High oilfield on the western coast, and because of the declines in international oil prices. The favorable energy environment was further supported by a series of good annual monsoon rains until the drought of 1987, which was readily offset by the economic recovery and growth of the 1980s and the buffer food stocks that had accumulated. However, the political and security implications of India's spreading nuclear nexus are troubling.

Nor has the long-term energy question been resolved for India. Its known oil reserves are low, probably not sufficient beyond the end of this century, and greater dependence on external oil in the future cannot be discounted. While India has substantial reserves of coal, there are various technical, economic, and political limitations on the extensive use of coal in the industrial, transportation, and household sectors, and the generation of hydroelectric power has its own set of problems and limitations. As a consequence, the promise of nuclear energy continues to draw the attention of India's economic planners and policymakers.

Within the triangular linkage of India's energy, security, and development policies, the interactions are complex and varied. For analytical convenience, two basic aspects of the problem initially will be examined here: (1) the manner in which conventional energy issues affect Indian development and security policies, and (2) the implications of India's nuclear energy policies for its development and security planning. The first level of analysis will examine Indian policy responses to international and domestic energy shortages and pricing policies, while the second level will focus on the nature of nuclear energy policy-making in India and its effects on domestic development and regional security.

CONVENTIONAL ENERGY

From the standpoint of conventional energy development, the sources and patterns of energy utilization in India—especially the degree of dependence on external sources of energy—were the subject of several official inquiries conducted by the Ministry of Energy and the Planning Commission following the international energy crisis of the 1970s. They sought to determine India's total energy needs until the year 2000, and they adopted interim as well as long-term policies to resolve those needs.

From the security standpoint, the procurement of conventional energy raises problems of both internal and external nature. Domestic political violence is caused directly by shortages, supply dislocations, or high prices of conventional energy and indirectly by energy-related economic failures. This factor was particularly relevant in the case of oil procurements in the middle to late 1970s. Coal-related internal security problems arise primarily from supply dislocations caused by strikes by railway workers or coal miners. Problems of external security, similarly, may be caused by supply dislocations arising from wars in the Middle East or by OPEC supply and pricing policies, as demonstrated by the 1973 Arab-Israeli war and subsequent OPEC pricing policies, and the 1980-88 Iran-Iraq war, which caused the partial loss of supply from these two countries. Indirectly, the various arms buildups in the Middle East have had some relevance for the strategic military balance in South Asia.

India has three conventional sources of commercial and industrial energy: oil, coal, and hydroelectric power. Oil of various kinds and oil products are used primarily in the transportation, petrochemical, and household sectors of the economy. The other two energy sources—coal-fired thermal and hydroelectric plants—provide more than 95 percent of electricity generation in India. Coal is also used in locomotives in some sections of the Indian railway network. In the rural economy, where more than 75 percent of the Indian population lives, there is still considerable use of firewood and agricultural and animal waste. The extensive use of cow dung as fuel in the rural areas has led to the setting up of several biogas plants under the Indian five-year plans.[1] The number of such plants grew from 6,900 in 1973 to 610,000 in 1986. In addition to biogas, the government of India has been exploring the potential use of geothermal, tidal, and solar power.

With a population of about 850 million in 1989, India's per capita consumption of petroleum is among the lowest in the world; among nations with sizable populations, only Bangladesh consumes less petroleum per capita. On the supply side, the Planning Commission's energy assessment in 1979 estimated that India had about 295 million metric tons (MT) of net recoverable oil reserves, of which 109 million MT (37%) was onshore and 186 million MT (63%) offshore. More recent estimates place total Indian oil reserves at about 471 million MT.[2] Gas reserves were estimated at 90 billion cubic meters, of which 67 billion cubic meters (74 percent) was onshore and 23 billion cubic meters (26 percent) offshore.[3] The oil reserves are clearly small compared with those of some of the major oil-producing countries of the world: Saudi Arabia (23 billion MT), Mexico (4 billion), the United Kingdom (2 billion), and Indonesia (1 billion). India's proven reserves are hardly sufficient to sustain its oil needs over the next decade.

Domestic production of petroleum in India rose from 11.9 million MT in 1978 to an estimated 17 million in 1980, while imports held steady at 14.4 million MT in 1978 and 15 million in 1981. The 1978 Indian production and import figures compare with 51 million MT of domestic production and 67 million MT of imports for the United Kingdom, 104 million MT of production and no imports for China, and 8 million MT of domestic production and 43 million MT of imports for Brazil.[4] By fiscal year 1984-85, India consumed 38.8 million MT of crude oil, of which 29 million were produced at home and 13.6 million imported. This move toward self-sufficiency is all the more creditable since India also exported 6.5 million MT of crude oil, for net imports of only 7.2 million.[5]

Oil and security. The increased demand for crude oil in India since 1978 has been met almost exclusively through stepped-up domestic production. Nevertheless, imports remain relatively substantial as demand steadily increases, and the disrupting effects of the international oil crisis that started in 1973 are not likely to be easily forgotten. Historically, the variations in and uncertainties of oil supply and prices have affected the Indian economy in two ways: the impact on India's economic development planning in general and the more specific consequences for certain sectors of the economy that are directly or indirectly dependent on oil.

At the broader level, the escalating oil prices between 1973 and 1981 drained India's foreign exchange reserves and created adverse trade balances. India's oil import bill rose from $264 million in 1972 to $1.4 billion in 1975 and $1.8 billion in 1978.[6] A series of price hikes after 1978 raised international prices to $31 to $42 per barrel and caused India's 1981 oil import bill to climb to approximately $7.5 billion,[7] consuming 80 percent of India's export earnings and constituting almost 50 percent of the total import bill.[8] (In comparison, the oil import bill had been less than 10 percent of the export earnings in 1973 and only 25 percent in 1974, immediately before and after the Arab-Israeli war, when OPEC prices were raised from approximately $4 per barrel to $12 per barrel.)[9] By 1986, however, imported oil costs fell to only 19 percent of India's export earnings, indicating that the situation had been stabilized.[10]

After the initial jump in oil prices in 1974, government planners had assumed a certain degree of international stabilization of both oil supplies and prices, with steady incremental increases of about 10 percent annually. For instance, the outline of the Sixth Plan, drafted in 1978, projected a demand of 36 million MT in 1982-83, of which 18 million MT was expected to be imported.[11] The anticipated price then was $24 per barrel ($175 per MT), for a total import bill of $3.15 billion. Because of the unexpected OPEC price hikes after 1978, this projection turned out to be less than half the actual 1981 import bill of $7.5 billion—a level that had not been anticipated until 1988.

OPEC oil prices thus had adverse effects on India's economic development plans during this period, despite the low per capita consumption. Indeed, the Fifth Five Year Plan, launched in 1974, was terminated in disarray early in 1978, although it should be noted that other economic and political factors also contributed to this situation. Between 1978 and 1980, the government of India switched to annual plans while it took stock of economic conditions. The Sixth Five Year Plan was then launched in 1980, only two years after the end of the Fifth Plan. During this period, the increasing diversion of critical foreign exchange reserves to relatively small amounts of oil imports deprived key government projects and kept private industrial development from importing necessary machinery and parts. Such withdrawals of potential investments in the public and private sectors tend to produce reverse multiplier effects on the economy as a whole.

Specifically, accelerating costs of imported oil affected the transportation, petrochemical, agricultural, and household sectors of the economy. Transportation typically takes nearly 50 percent of the available refinery products, with the Indian Railways alone consuming about 20 percent. The domestic household sector annually consumes 30 percent, mainly in the form of kerosene for cooking. The petrochemical industrial sector absorbs about 13 percent, mainly for the production of fertilizers, basic polymers and organic chemicals, plastics, manmade fibers, drugs, detergents, paints, pigments, and dyes.[12] The agricultural sector accounts for about 5 percent, principally to fuel tractors and diesel pumps for irrigation. Only 1 percent of oil is used to generate electric power.

Several extended economic and political disruptions tend to occur in India along with dislocations in oil supplies and higher oil prices. Higher

operating costs in the petrochemical industries and transport industry
tend to increase sharply the prices of a range of intermediate and consumer products, as well as urban transport services, which affect the spending
power of the more politically vocal middle classes. Higher prices for
kerosene are a hardship for urban and rural lower middle classes, who use
this basic fuel for home cooking. Consequently, while the household
sector may consume only one-third of available oil, higher oil prices draw
adverse political reactions from the active, vocal section of the Indian
population concentrated in cities and towns. Higher oil prices and disruptions in supplies may limit both fertilizer production and the availability of
diesel oil for irrigation pumps and tractors, thus curtailing agricultural
production as a whole, with ominous political repercussions.

The disruptive effects of the oil crisis on agricultural production have
been checked to some extent by favorable monsoon rains in most years
since 1970. The exceptions were 1974, when there was only 0.1 percent
growth in the Indian GNP, and 1979, when it declined by 4.5 percent.[13]
The monsoon failure of 1973, in conjunction with the broader effects of the
oil crisis, pushed the 1974 inflation rate to an unprecedented 27 percent,
causing a general economic failure.[14] The large-scale political demonstra-
tions against the Congress government of Indira Gandhi that followed in
early 1975 led eventually to the declaration of an "emergency" in June and
a suspension of democratic rights.[15] The partial monsoon failure in 1979,
followed by steep increases in oil prices during 1980 and 1981, produced
similar but less destabilizing economic and political conditions.

Another economic and political aspect of the oil crisis in India was the
discrepancy in the prices of imported oil and domestic oil. For instance,
the domestic oil price until 1981 was pegged at $4.50 a barrel for the
onshore variety and $6 per barrel for the offshore.[16] In 1981, prices were
standardized at $18 per barrel for both onshore and offshore oil—still much
lower than international oil prices. The Indian government, in other
words, was heavily subsidizing the overall cost of oil for the Indian consumer during the height of the crisis period, despite the exorbitant import
price and high taxes on petroleum and petroleum products.

The pricing policy for domestic oil in the early 1980s led to major
political disturbances in the state of Assam, where the bulk of domestic
onshore oil is produced. Political agitators in Assam demanded that their
state be compensated for its production at OPEC oil prices. Demonstra-
tions in 1981 forced the closure of the 1 million MT Bongaigon refinery and
caused intermittent shutdowns of the 1 million MT Gauhati refinery and
the smaller refinery at Digboi. Subsequently, similar public demands
were voiced in the state of Bihar, where oil is refined, leading to the
temporary closure of the Barauni refinery.

Along with the internal security problems in India caused by the oil crisis during the 1973-82 decade, the arms buildup in the Middle East, funded by the accumulation of petrodollars, posed two potential threats to India. There were fears both that Middle East arms purchases might eventually find their way into Pakistan during another subcontinental war and that some of the surplus petrodollars might be used to finance and assist Pakistan in developing a nuclear weapons capability. In the 1960s, Iran and Jordan had transferred a few F-86 Sabre combat aircraft to Pakistan,[17] and the practice of sending Pakistani military training missions to Islamic countries such as Libya, Saudi Arabia, and the United Arab Emirates suggested that Pakistan might acquire familiarity in the deployment of weapons acquired by the oil-rich states of the region. The nuclear threat—the prospect that Pakistan might develop the "Islamic bomb"— was allegedly being financed either directly or indirectly by Libya and Saudi Arabia, with uranium supplies coming from Libya via the Muslim state of Niger.[18]

Neither of these two prospects have so far been realized. Indeed, the fall in international oil prices and the declining revenues of the OPEC countries have begun to sever Pakistan's military ties with the oil-producing Arab states. Pakistani army divisions in Saudi Arabia have been withdrawn, and other military training personnel in the Gulf states are being cut back. Economic declines in the Gulf sheikdoms have also ended the employment of a large immigrant labor force from India and Pakistan. The return of this labor force to the subcontinent has produced regional social and economic pressures. In India, much of the Gulf labor had come from the state of Kerala, ruled by the Communist Party-Marxist (CPM) either directly or through coalitions with other left-leaning parties, including the Congress. Property purchases by this wealthy returning emigré group have caused real estate prices to rise rapidly, and this new special landowning class has produced resentment among traditional landowners and agrarian workers.

Coal, the main source of commercial energy in India, provides almost 65 percent of total electricity generated in the industrial sector. Coal reserves in India are substantial, projected to last another 100 years at compounded rates of consumption. In 1985, the Geological Survey of India assessed total coal reserves in India at 156 billion metric tons for coal seams of 0.5 meters and thicker recoverable from depths to 1,200 meters.[19]

Despite such large proven reserves, the productivity of operations under the government-owned corporation Coal India Limited and its subdivisions is low when compared with Western standards. The average output per man-shift for Coal India is 0.79 MT in underground mines, considerably lower than the range of about 2.5 to 4 MT in Western Europe and 8 to 12 in the United States, Australia, and South Africa.[20] The figures reflect the low level of mechanization in India, where cheap labor is abundant. Indian productivity, however, compares favorably with that of China, where output per man-shift is about 0.5 MT for underground mines and 1.6 for open-pit mines. Although India's Sixth Five Year Plan (1980-85) had allotted an average annual investment of 8 billion rupees in the coal industry, compared with only 2.13 billion during the previous four years, the mechanization process is likely to remain slow because of political and social pressures to absorb the vast pool of labor in India.

The energy crunch in India during the 1970s as a result of the international oil crisis led to greater economic rationalization in the coal industry. Until 1981, coal prices in India were set below the cost of production, and the pricing system excluded the rate of return and depreciation from estimates of the cost of production. Wages and salaries, for instance, which generally accounted for 70 percent of the cost of production, were not adequately assessed and incorporated when determining coal prices. Under this subsidized system of accounting, Coal India Limited posted a loss of Rs 7 billion between 1974 and 1979.

After 1981, pricing was linked to the cost of production, and the government allowed Coal India to set prices that would earn a return on investment. The modest profit of Rs. 300 million in fiscal year 1981-82 was expected to grow to Rs. 1 billion the following year.[21] The Sixth Five Year Plan (1980-85) paid greater attention to the needs of the coal industry. Greater emphasis was placed on coal exploration, so as to raises the annual rate of production from 119 million MT in 1980-81 to 155 million MT in 1984-85 and 161 million in 1985-86.[22] By way of comparison, annual coal production was 33 million MT in 1950-51, 56 million in 1960-61, and 76 million in 1970-71. Much of the acceleration in coal production was accomplished through greater mechanization in coal drilling, the reconstruction of older coal mines, and the development of new mines.

Assessing the total potential of hydroelectric resources in India is difficult because the flows of Indian rivers vary widely, ranging from thousands of cusecs in the monsoon season to a few cusecs in the dry season.[23] Consequently, estimates of hydroelectric potential in India usually amount to identifying specific targets in each river basin with the potential for hydroelectric power generation that is both technically feasible and economically viable. Based on these criteria, the first systematic survey of the Central Water and Power Commission, conducted between 1953 and 1960, identified 260 possible targets with a potential annual energy generation of about 221 tetrawatt-hours (TWH). The estimate was revised in 1978 to 396 TWH. However, the installed hydroelectric generating capacity in India in 1978 was only 39 TWH, or just 10 percent of estimated potential, a share that changed little during the next decade. Moreover, the utilization of the available potential has not been uniform among the different regions.

The Planning Commission's energy report also assessed the potential of India's hydroelectric capacity in terms of megawatts (Mw) at a specified load factor. In other countries, a load factor of 60 percent tends to represent the potential in terms of capacity, but in India the average load factor of hydroelectric plants is about 42 percent. At this load factor, the total power potential was estimated at 100,000 Mw, while the corresponding total installed capacity in 1978-79 was only 10,830 Mw. The Planning Commission noted that the utilization of only 10 percent of India's hydroelectric potential amounted to considerable wastage; renewable hydroelectric potential is being lost each year because of the failure to harness India's rivers.

If we assume that energy security is best achieved in India through the development of conventional sources rather than nuclear energy, then both coal and hydroelectric power will continue to dominate. If this is the case, certain internal security problems may arise. The problem with coal is that it is located almost exclusively in the volatile northeast sector of the country. Coal production to supply power plants throughout the country is vulnerable to disruption, especially in West Bengal, where violent labor strikes and general political unrest have been extensive.

Similar complications could interrupt the transportation and distribution of coal throughout India by railway, since the All-India Railwaymen's Federation is one of the largest and most powerful labor unions in India, with a membership of about 2 million. A nationwide strike would not only bring all railway passenger and goods traffic to a halt, but would also paralyze coal-fired power plants and thereby industrial activity throughout India. The seriousness of this internal security threat is probably best exemplified by the crisis of 1974, when the All-India Railwaymen's Federation successfully launched a nationwide strike of its nearly 2 million members.[24] With most power plants carrying less than two weeks of coal supplies, power had to be severely rationed in major industrial cities that were dependent on electricity from coal thermal plants.

Energy supply dislocations due to the 1974 railway strike intensified the energy crisis induced by OPEC supply and pricing policies, causing a severe economic crisis in fiscal year 1974-75. The economic growth rate fell to less than 2 percent, and inflation rose from 8 percent in 1972-73 to 16 percent in 1973-74 and to 29 percent in 1974-75.[25] The economic crisis, in turn, produced widespread civilian violence that threatened to undermine the stability of the country. Unable to control the growing internal political chaos in 1975, Prime Minister Indira Gandhi invoked the emergency powers of the constitution on an unprecedented nationwide scale. Under these powers, the democratic process in India and the fundamental rights of citizens were suspended. To be sure, it would be unfair to attribute the national political crisis of 1975 to the railway-induced coal shortages and the OPEC-induced oil price shocks of 1974, but these energy happenings were among the internal security problems of the mid-1970s.

There have been no major internal security-related disruptions of hydroelectric power, such as sabotage of power plants during violent industrial unrest or rioting on a communal basis (religious, linguistic, caste, or tribal). However, India has failed to harness such power in the politically turbulent northeast sector, where the potential for such energy development is among the highest. Less than 0.5 percent of the hydroelectric potential has been developed in the northeast,[26] a fraction of the 10 percent nationwide development of hydroelectric potential.

The northeast sector of India, which comprises West Bengal and Assam, and the tribal states of Nagaland, Mizoram, Meghalaya, Manipur, and Tripura, has been perennially riddled with industrial unrest, violent communist movements, religious and linguistic conflict, and tribal guerrilla warfare seeking to establish independent homelands. Since this sector is the site of two major river basins—those of the Ganges and Brahmaputra and their tributaries—an argument might be made that the failure to develop the immense hydroelectric potential even moderately is due to political turbulence.

The availability of alternative energy sources may help explain the failure to develop the hydroelectric potential in the northeast, which is also rich in coal and oil resources. However, coal is more expensive than hydroelectric power for electricity generation in India, even in the northeast, and oil in India is intended almost exclusively for the transportation and the petrochemical industries. Given these circumstances, hydroelectric energy would appear to be the logical choice for the region.

The failure to tap the hydroelectric potential may arise also from "planning" assessments—that is, technical difficulties foreseen in harnessing the rivers, or from the lack of financial resources to invest in the region. However, government support for the development of cheap hydroelectric power would attract private industrial investment and contribute to economic growth in the region. Despite good reasons for greater central government investments in the region, the failure to do so would suggest that political instability may be the primary reason why government planners have avoided exploitation of its hydroelectric potential.

NUCLEAR ENERGY

Whatever the drawbacks, coal, thermal, and hydroelectric power plants will continue to generate the bulk of electricity in India for several decades to come. Nevertheless, the growth in thermal and hydroelectric generating capacity is not expected to keep pace with industrial and urban demand. Energy projections in India beyond the year 2000 suggest serious shortfalls of up to 20 percent. That crucial gap cannot be filled by increasing oil production, since, domestic oil reserves are too limited to meet industrial demand and, in any case, will be depleted in fifteen to twenty years. All future oil imports are expected to be absorbed by the petrochemical, household, and transportation sectors of the economy. Likewise, as noted earlier, the concentrated geographic location of coal and the vulnerability of railway transportation mean that coal cannot guarantee continuous energy supply throughout India. Under these circumstances, Indian planners and politicians perceive nuclear power as critical.

Leading Indian nuclear scientists and economic planners are convinced that nuclear power is the only long-term solution to India's energy needs. The following statement by Raja Ramana, director of the Bhabha Atomic Research Center and secretary of India's national Department of Atomic Energy, is reflective of the views of the nuclear scientific community in India: "Looking at it [the future of nuclear energy] with the experience of the past and the terrifying energy problems of the future, I can think of no other source of energy that has been discovered to date except nuclear energy, which can solve the energy problems of this country during the next 25 years and beyond. If I do not make the case now and point out to the urgency of accepting its inevitability, I will have done a great disservice."[27]

Thus, despite the relatively small contribution of nuclear power to the Indian economy, it plays a critical role. The current and projected uses of nuclear energy, however, carry political, economic, and security implications. The basic question remains: Are nuclear energy programs in India intended primarily to maintain India's nuclear weapons option?

The Department of Atomic Energy was set up in 1954 as the executive agency of the Indian Atomic Energy Commission, which had been established six years earlier. Technological and economic self-sufficiency was the key words for the atomic energy program as mapped out in 1947, soon after independence, by Homi J. Bhabha, first director of the Department of Atomic Energy.

Bhabha's long-term plan envisaged three stages of development.[28] The first stage was to establish natural uranium–fueled pressurized heavy water reactors based on the Canadian design known as CANDU. Heavy water reactor (HWR) technology was preferred to the U.S.-designed light water reactor (LWR) technology because HWRs did not require enriched uranium but could eventually utilize thorium. India has only about 52,000 metric tons of commercially exploitable uranium reserves, while thorium reserves total about 320,000 metric tons. HWRs furthermore, were perceived to be economically more efficient than LWR systems and had the added military advantage of producing plutonium—the fissile weapons-grade material—as a byproduct. Except for the Tarapur Atomic Power Plants, which use enriched uranium, all existing nuclear plants, as well as all those planned for the near future, use natural uranium and pressurized heavy water. However, there have been some recent moves by the government of India to obtain two Soviet LWRs of 1000 MWe each, with the possibility of later building similar reactors in India.[29]

The second-stage nuclear reactors were to be fueled by plutonium as well as uranium-233 derived from thorium placed in the blanket. The third stage of nuclear development was expected to be based on thorium-fueled fast breeder reactors, which would lead to a self-sustaining thorium-uranium fuel cycle. Since India has vast resources of thorium, the nuclear energy program was expected to become commercially viable beyond the year 2000. A fast breeder test reactor based on a French design has already been constructed at the Research Reactor Center at Kalpakkam, Madras. Nearly all the major components of the reactor were fabricated in India.

Although the Bhabha plan (later elaborated by the next director, Vikram Sarabhai) did not recommend LWRs, successful negotiations with the United States led to an agreement in 1963 for the construction and installation of two LWRs at Tarapur in Bombay, known as TAPP-1 and -2. The two LWRs were provided to India on a turnkey basis by the General Electric Company of the United States. The Indian government, private Indian industry, and labor provided 30 percent participation in setting up TAPP,[30] which has been operating since 1969. The mainstream line of nuclear development in India, however, was based on the Canadian-designed HWRs, starting with construction of the Rajasthan Atomic Power Plant. The two heavy-water units of 200 MWe each at Rajasthan (RAPP-1 and -2) were supplied by the Atomic Energy Commission of Canada. The Rajasthan plants have been operating since 1973 with 70 percent participation and funding by Indian industry and labor. Since then, two HWRs with capacities of 220 MWe each have been indigenously designed and constructed at Kalpakkam near Madras (MAPP-1 and -2), and two similar reactors at Narora in Uttar Pradesh (NAPP-1 and -2). Four more similar reactors are under construction at Kakrapar in Gujerat and at Kaiga in Karnataka. An additional six such reactors are to be developed and installed by the year 2000. Almost all of the development will have been done exclusively by Indian technology, industry, and labor.

It is not clear that all of these projects will be completed on schedule, and the projections may be highly optimistic. Only the Tarapur, Rajasthan, and Kalpakkam plants were generating power for commercial purposes by 1988, although the Narora plants were expected to be in operation by 1989. Thus, while government projections indicate an installed nuclear generating capacity of 8,620 MWe by the year 1991 and 10,000 MWe by the year 2000, other parliamentary and ad hoc forecasts have been less optimistic.[31] For example, the Estimates Committee in 1978 forecast 6,000 MWe by 1991, and a 1979 ad hoc report of the Working Group on Energy Policy of the Planning Commission predicted the installation of only 5,000 MWe of nuclear power by the early 1990s.[32] Failure to approach these goals led to the immediate purchase of two 1,000-MW light water reactors from the Soviet Union in 1988. These two reactors were to be installed in the southern region of Tamil Nadu to meet power shortages in both that state and Kerala.

Whatever the discrepancies in projections of installed nuclear power over the next two decades, comparative cost estimates by the pronuclear power lobby in India suggest that this source of energy may have a cost advantage over coal-fired thermal plants, despite indications to the contrary. According to Raja Ramana, secretary of the Department of Atomic Energy, the apparent higher cost of nuclear power over thermal power arises from the nature of cost accounting applied to these two sources of energy.[33]

An assessment of the average cost of nuclear-generated electricity includes the capital outlay in areas like the production of heavy water and the setting up of a means of controlling the nuclear fuel cycle. On the other hand, according to Raja Ramana, the extensive costs of developing coal mines and transporting the coal to power plants throughout India are not included in the assessment of the average cost of coal-generated electricity, mainly because the development of coal mines, the construction of railway lines, and the acquisition of wagons and locomotives are not undertaken exclusively for generating electricity from coal-fired thermal plants. Ramana said, "The point usually overlooked is that for each coal-fired power station replaced by a nuclear power station, the need to develop coal mines and to organize transportation facilities will be proportionately reduced, and thus, the capital outlay required for these purposes would be significantly reduced."[34]

Allegations of military motives behind India's nuclear energy program are based on the security pressures, especially nuclear threats, perceived to stem from China and Pakistan, and on the questionable commercial viability of nuclear power vis-á-vis coal-fired thermal and hydroelectric power in India.

Since China exploded its first atomic bomb in October 1964, there has been debate over whether India should acquire its own nuclear weapons to counter the Chinese threat. In the policy arena, there are three basic alternatives: maintaining the nuclear weapons "option" (India's current policy), exercising that option, or rejecting it. The nuclear weapons option leaves open the possibility of embarking on a military program. The primary and specific purpose of this policy has been to increase the pressure on the superpowers to provide credible nuclear guarantees against the nuclear threat from China and potentially from Pakistan so as to improve the climate of regional security. The secondary and more general purpose is to exert pressure on the nuclear "haves" to reduce and eventually eliminate their own nuclear stockpiles so as to improve the climate of global security. Maintaining the military option justifies India's nuclear energy programs even if the outlay substantially exceeds the costs of developing alternative sources of energy.

If India should decide to exercise its nuclear option by embarking on the "dedicated path" toward weapons production, the country's policy would be secondary or even incidental to the need to acquire weapons for security purposes. The strategic implications of a weapons program— quite different from the case of merely maintaining the weapons option— would trigger a three-way nuclear arms race among India, Pakistan, and China. Pakistan would almost certainly respond with an overt weapons program of its own, while some of the Chinese missiles now directed against the Soviet Union would almost certainly be targeted instead at India. The net result might be less security for India at a higher economic cost. The first Indian atomic test of May 1974 appeared to have set this process in motion. Although India quickly reverted to the policy of maintaining the nuclear weapons option after the outcry in the United States and Canada, the 1974 test increased the determination in Pakistan to acquire the capability to produce nuclear weapons.

India's third alternative would be to sign the Non-Proliferation Treaty and give up its nuclear weapons option. While the technological capability would persist through the continuation of the nuclear energy program (despite International Atomic Energy Agency inspection and other safeguards under the Non-Proliferation Treaty), this course of action would at least dampen the Indian temptation to acquire nuclear weapons and lessen the probability of nuclear proliferation in South Asia. The establishment of a nuclear-free zone in South Asia would in fact serve as a model for similar zones in other regions of the world. This third approach would also open greater international technological cooperation for India in the peaceful uses of the atom, including nuclear energy.

India's current policy of maintaining the nuclear weapons option serves the needs of both defense and development programs. By wielding the military option, India has avoided the vulnerability of a unilateral renunciation of nuclear weapons on the one hand, and the security risks of embarking on nuclear arms races with China and Pakistan on the other. Indeed, thus far this approach has threaded an effective course between the extremes of action (a nuclear weapons program) and inaction (unilateral renunciation). At the same time, despite some American and Canadian efforts to stall the nuclear energy program, a degree of cooperation is likely to persist to avoid pushing India toward the dedicated path of nuclear weapons production.

Maintaining the military option clearly suggests deliberate ambiguity about the purpose of India's civilian nuclear energy program. External allegations of military motives rest mainly on the questionable commercial viability of nuclear power as compared with coal-fired thermal and hydroelectric power in India. Cost estimates in the United States, West Germany, Japan, France, and other Western nations suggest that the earlier optimism about LWRs, HWRs, and fast breeder reactors as a source of energy was misplaced. The scarcity and high cost of uranium resources, the high capital cost and growing technical complexities in nuclear plant design and operation, fears of plant meltdowns and other radioactive hazards, and the risk of tempting terrorists with critical targets—all are claimed to have greatly reduced the prospects for commercial nuclear energy.

The Indian counterargument is that pessimistic claims regarding the future of nuclear power are motivated by Western efforts to curb nuclear proliferation rather than being objective assessments. From this point of view, for instance, it was no coincidence that President Jimmy Carter followed both a domestic policy of curbing the growth of nuclear power and an international policy of curbing nuclear weapons proliferation.[35] Any demonstration of the viability of nuclear power in the United States would feed a worldwide demand for nuclear reactors. Eventually, control over the nuclear fuel cycle in Third World countries would set the stage for production.

In India, however, a nuclear weapons program would be much less cost-effective than current conventional defenses. In India's case, there appears to be no relevance to France's argument under President Charles De Gaulle in the early 1960s that a nuclear weapons program would reduce the overall cost of defense by reducing the need for conventional forces. In India, a defense program is often judged by whether it is labor-intensive or capital-intensive. In a nation with an abundance of cheap labor, capital-intensive proposals such as those of the Indian air force and Indian navy require a convincing strategic justification. The labor-intensive programs of the Indian army have generally consumed about two-thirds of the annual funds allocated to Indian defense.[36] Pakistan has responded with similar types of military planning, setting in motion a self-perpetuating strategic justification for emphasizing conventional army programs.

In the light of these economic and strategic rationales, a nuclear weapons program would appear unattractive, even self-defeating. It would employ only a small scientific community in India (in 1981-82, the Department of Atomic Energy employed 19,158 scientific and technical personnel and 12,498 administrative and auxiliary staff).[37] It would trigger strategic nuclear responses in Pakistan and China, setting off a costly and dangerous arms race in South Asia.

In short, there are both civilian and military purposes underlying India's nuclear energy program. Maintaining the weapons option serves to justify outlays for nuclear power whatever the current commercial viability of this source of energy. At the same time, nuclear power is desirable in its own right, since it is expected to cover potential shortfalls between India's total demand and the conventional sources of energy, and nuclear energy appears all the more attractive in the face of OPEC's volatile pricing and supply policies of the recent past. The civilian pursuit of nuclear energy would give India a basis for embarking on nuclear weapons development if necessary, on the grounds that the cost would be only incrementally higher than for a nuclear energy program. While it might be difficult to justify either nuclear energy development or a nuclear weapons push on their own, taken together the two initiatives provide a mutually reinforcing strategic-economic justification.

CONCLUSIONS AND PROSPECTS

The overall degree of development in India may be measured by the growth in electric power generation for the industrial, agricultural, and household sectors, which relies almost exclusively on coal-fired thermal, hydroelectric, and nuclear plants. While oil is critical in the transportation and petrochemical industrial sectors and, as kerosene, for cooking in the household sector, it is not used much for power generation except in the agricultural sector. The pace of growth in electricity generation has steadily improved since the Third Five Year Plan (1961-66), with the annual rate of growth nearly doubling by the 1980s. (See table 2.1.)

While the major role in electricity generation is played by coal-fired thermal plants, the importance of hydroelectric power is rapidly increasing and the planned shift in this direction reflects the abundance of untapped energy that may be obtained from India's rivers. Indeed, much of northern and northeastern India may be serviced by hydroelectric power if the waters of the mighty Ganges and Brahmaputra rivers and their tributaries are properly harnessed. Nuclear power remains insignificant in

Table 2.1.   Growth of power-generating capacity in India
             (Megawatts)

	Coal	Hydroelectric	Nuclear	Total
1961-68				
Total added	19,281 (62%)	11,386 (36%)	640 (2%)	31,307
Avg. annual addition	1,134	670	38	1,842
1980-85				
Total added	10,898 (77%)	2,873 (20%)	455 (3%)	14,226
Avg. annual addition	2,180	575	91	2,845
1985-90				
Planned additions	15,999 (72%)	5,541 (25%)	705 (3%)	22,245
Planned avg. ann. add.	3,200	1,108	141	4,449
Total installed				
capacity, 1986	29,856 (64%)	15,477 (33%)	1,270 (3%)	46,603

Source: Figures derived and calculated from Ministry of Information and Broadcasting, *India: 1986*, New Delhi, 1987, p. 396.
*This period of years was chosen for comparative purposes only.

terms of total power generated, partly because of serious teething problems in the nuclear industry that are expected to be overcome by 1990. In the next decade, nuclear power is expected to contribute about 10 percent of the total power generated in India.

More than 15 years after the onset of the international oil crisis in 1973, the energy crisis in India now appears over or is subdued, as elsewhere in the world. The oversupply of oil on the international market from the mid-1980s onward, arising from the entry of Mexico into the export market, the production of North Sea oil, and the completion of the Alaskan oil pipeline in the United States, has sowed dissent among the OPEC nations regarding production quotas and controls on prices. These trends at times over the past 15 years have brought international oil prices down to about $10-14 per barrel, roughly the level that preceded the crisis in 1974.[38] Although the 1980-88 Iran-Iraq war of attrition cut off much of the oil supplies available from these two countries, India managed to maintain modest imports from both these sources, meanwhile diversifying its suppliers and increasing domestic production.

Domestic oil exploration and production, in particular, has contributed substantially toward relieving the import burden. There has been only a marginal annual increase in Indian oil imports since 1973, with the bulk of Indian needs now being supplied domestically. Oil production in India increased from 7 million MT in 1973 to 14 million in 1980 and to 29 million in the fiscal year 1984-85.[39] The 1984-85 domestic production figure represents almost 75 percent of the total oil procurement for that period. The improvement in domestic production is likely to continue under the government's new policy of seeking greater participation from Western multinational oil corporations for oil exploration in India.

Although there is the potential for supply dislocations in coal arising from strikes by coal miners and railway workers, there have been no major problems in this respect in recent years. India continues to depend on railway transportation to keep coal-fired thermal plants in operation throughout the country, as highlighted by recent increases. Coal supplies to power stations throughout India during the fiscal year 1981-82 totaled 40 million MT, about 20 percent higher than in the previous year. There were 4,365 railway wagons available in January 1982 for coal transportation, up 12 percent the previous year.[40]

The incorporation of the public sector National Thermal Power Corporation (NTPC) and the National Hydroelectric Power Corporation (NHPC) in 1975 was intended to speed up the installation of new conventional power plants. The NTPC is currently in the process of setting up seven coal-fired super power stations with a total installed capacity of 10,900 MW, at Singrauli and Rihand (Uttar Pradesh), Korba and Vindhyachal (Madhya Pradesh), Farraka (West Bengal), Kahalgaon (Bihar), and Ramagundam (Andhra Pradesh). Similarly, the NHPC is in the process of implementing the Salal and Dulhasti hydroelectric projects (in Jammu and Kashmir), as well as hydroelectric plants at Koel Karo (in Bihar), Chamera (in Himachal Pradesh), and Tanakpur (in Uttar Pradesh). However, in spite of fresh efforts to harness India's rivers, hydroelectric power remains inadequately exploited, and a clean and renewable source of energy is wasted.

Since the mid-1980s, there has been a softening of international efforts to curb India's nuclear energy program, especially after the Reagan administration chose to play down antiproliferation in favor of wooing Pakistan to fight the Soviet military presence in Afghanistan. On the other hand, there appears to be no overt effort by India at present to acquire nuclear weapons. Efforts by the U.S. Congress in the late 1970s to block the supply of enriched uranium for the Tarapur nuclear power plant in Bombay, because of India's refusal to sign the NPT, have been partly offset through an arrangement negotiated by the Carter administration with France to provide the enriched uranium to India. Meanwhile, various technical difficulties continue to delay India's development of nuclear energy.

The main setback has been the denial of heavy water by Canada for the CANDU power reactors after India's efforts to obtain heavy water from its plants proved to be a dismal failure. The heavy water plants built in India have functioned at less than 10 percent of capacity, and only three of the five plants were operative by the mid-1980s. Lack of heavy water has at times idled the Rajasthan and Madras nuclear reactors. To keep the Rajasthan reactor in operation, India turned to the Soviet Union for imports of heavy water, and this external dependence has persisted. In 1988, press reports claimed that India had also obtained heavy water from Norway.[41]

The more uncertain and disturbing aspects of the Indian energy problem for the future are the political and security-related byproducts. Even though the general energy supply situation has improved, certain political and security legacies of the 1970s remain. In the first place, the oil crisis reinforced the Indian commitment to nuclear energy despite increasing international and domestic constraints. Since nuclear energy objectives are not likely to be discarded—indeed they might have persisted even without the oil crisis—India will stay a short step away from a major nuclear weapons capability at an acceptable economic cost. This potential initially set Pakistan on the "dedicated path" toward acquiring nuclear weapons capability and subsequently toward an imitative policy of nuclear energy initiatives that may be diverted to weapon purposes at short notice. An overt or covert Indo-Pakistani nuclear arms race will have implications for the Middle East as well, as both countries compete for economic favors from this region in exchange for nuclear technology and equipment. Such an eventuality would accelerate nuclear weapons proliferation and increase the risks of nuclear war in the region.[42]

In the second place, the glut of petrodollars in the Middle East resulted in a large-scale buildup of sophisticated conventional arms in the region, especially in Saudi Arabia. While this trend is likely to slow as OPEC oil revenues decline, there were fears in India at one time of Pakistani military links with Saudi Arabia and Libya that might advance Pakistani conventional and nuclear weapons capability. Following the Soviet invasion of Afghanistan in December 1979, the United States chose to arm Pakistan and Saudi Arabia. U.S. arms sales to Pakistan were partly financed by Saudi Arabia, one of the military links between the Middle East and South Asia. India's earlier fears of a Pakistani-Libyan link lay in the possible financing of a Pakistani nuclear weapons program. India's military counter-response to these trends was to acquire a substantial arms package from the Soviet Union that included MiG-23, MiG-25, and the advanced MiG-29 aircraft as well as T-72 tanks. India also procured the advanced Mirage-2000 combat aircraft. Meanwhile, any Pakistani development of nuclear weapons, whether overt or covert, with or without Saudi and Libyan assistance, is likely to trigger a major nuclear weapons initiative in India.

Indian economic and defense planners have translated the lessons of the international energy crisis into some pragmatic and successful policies in recent years. However, the persistence of old energy-related economic and security issues and the rise of new concerns in these areas will continue to test India's economic and security policies during the next decade.

NOTES

1. Se Raju G.C Thomas, "Energy Politics and Indian Security," *Pacific Affairs* 55, no. 1 (Spring 1982): p. 34. The 1986 figures were obtained from Ministry of Information and Broadcasting, *India: 1986* New Delhi, p. 414.

2. *The World in Figures, The Economist*, (London: Oxford University Press, 1984), p. 168.

3. Planning Commission, *Report of the Working Group on Energy Policy*, New Delhi. 1979, p. 20. Official estimates have tended to vary. According to the 1981-82 annual report of the Ministry of Energy, geological surveys indicated reserves of recoverable crude oil to be about 370 million metric tons, and reserves of the natural gas to be about 350 million cubic meters.

4. The 1978 figures are from *World Energy Supplies* (New York: United Nations Publications, 1979). The 1980 figures are from *India News* (Embassy of India, Washington, D.C.), Feb. 23, 1981. The comparative levels of proven reserves were derived from Ministry of Energy, *Report: 1979-80*, New Delhi, p. 37.

5. *India: 1986*, p. 407.

6. See Peter Sinai, *New Lamps for Old* (New Delhi: Indian Documentation Service, 1980), p. 6.

7. See *Hindustan Times*, Dec. 17, 1980; *Statesman*, Dec. 17, 1980.

8. *India News*, Feb. 9, 1981.

9. Figures obtained from Surjit S. Bhalla, "India's Closed Economy and World Inflation," in William R. Cline and Associates, *World Inflation and the Developing Countries* (Washington, D.C.: Brookings Institution, 1981), p. 160.

10. World Bank, *World Development Report: 1988* (New York: Oxford University Press, 1988), p. 222.

11. Figures obtained from Sinai, *New Lamps for Old*, p. 24.

12. See S.S. Khera, *Oil: Rich Man, Poor Man* (New Delhi: National Publishing House, 1979, pp. 216-17. S.S. Khera was formerly India's cabinet secretary.

13. Calculated from World Bank, *World Tables: 1987* (Washington, D.C.: World Bank, 1988), pp. 212-13.

14. See Bhalla, "India's Closed Economy," p. 137.

15. There are several studies of the origins and consequences of the "emergency" in India between 1975 and 1977. For a good discussion of the issues during the emergency itself, see Henry C. Hart, ed., *Indira Gandhi's India: A Political System Reappraised* (Boulder, Colorado: Westview Press, 1976).

16. See Swaminathan S. Aiyar, "The Hidden Oil Subsidy," *Hindustan Times*, Jan. 1, 1981. The domestic oil price of $18 per barrel at this time was reported by the *Economist*, July 18, 1981.

17. During the 1965 and 1971 Indo-Pakistani wars, Saudi Arabia, Libya, and some of the Gulf sheikdoms threatened to aid their coreligionists with arms. These threats were not carried out, but military ties between Pakistan and these countries during the oil crisis were cause for concern in India. Pakistan had trained the pilots of the United Arab

Emirates and Libya to fly the French-supplied Mirage-III and -V aircraft, which were also deployed by the Pakistan air force. For a discussion of these issues, see Raju G.C. Thomas, "Aircraft for the Indian Air Force," pp. 85-101.

18. For an Indian perspective of the issue at the time, see Major General D.K. Palit and P.K.S. Namboodiri, *Pakistan's Islamic Bomb* (New Delhi: Vikas Publishing House, 1979). For an American assessment, see Rodney W. Jones, *Nuclear Proliferation: Islam, the Bomb and South Asia*, Washington Papers, no. 82 (Beverly Hills, Calif.: Sage Publications, 1981). The allegation that Saudi Arabia, in addition to Libya, was helping to finance the development of a hydrogen bomb, was made by London's *Sunday Times* and given wide publicity in the Indian press. See the *Indian Express*, Jan. 19, 1981, and the editorial in the *Hindustan Times*, Jan. 20, 1981. See also report entitled, "Secret Pakistani Search for Nuclear Equipment," *Times of India*, Dec. 31, 1980.

19. *India: 1986*, p. 405. See also *Report of the Working Group on Energy Policy*, p. 18.

20. Jairam Ramesh, "Developing India's Coalfields: Rational Pricing Policy Needed," *Times of India*, April 14, 1982.

21. Jairam Ramesh in *Times of India*, April 14, 1982.

22. *India: 1986*, p. 432. See also speech by Vikram Mahajan, Minister of State for Energy, "Meeting India's Energy Needs," *Indian and Foreign Review*, August 1, 1981.

23. *Report of the Working Group on Energy Policy*, p. 22.

24. See Robert L. Hardgrave, *India: Government and Politics in a Developing Nation* (New York: Harcourt, Brace and Jovanovich, 1980), p. 132. See also Raju G.C. Thomas, "India's Nuclear and Space Programs: Defense or Development?" *World Politics*, 38, no. 2 (January 1986): pp. 315-42.

25. See Ministry of Information and Broadcasting, *India: 1981*, New Delhi, pp. 34, 166-68.

26. From *Report of the Working Group on Energy Policy*, pp. 21-22.

27. See Raja Ramana, "Inevitability of Atomic Energy in India's Power Programme," in Rajendra K. Pachauri, ed., *Energy Policy for India* (New Delhi: Macmillan, 1980), p. 221. See also observations of M.T. Srinivasan, "Where Should all the Energy Come From?" *Times of India*, Dec. 7, 1980. M.T. Srinivasan at the time was the director of the power projects engineering division of the Department of Atomic Energy.

28. From Department of Atomic Energy, *Report: 1980-81*, New Delhi, p. 4.

29. Foreign Broadcast Information Service, *JPRS Report: Nuclear Developments*, Sept. 2, 1988, p. 14.

30. See *Nuclear News* 24, no. 2 (1981). For an official assessment, see Ministry of Energy, *Estimates Committee, 1977-78*, 16th report, New Delhi, pp. 216-41. See also "Energy Politics and Indian Security," *Pacific Affairs* 55, no. 1 (Spring 1982), p. 41.

31. See Department of Atomic Energy, *Report: 1980-81*, p. 3.

32. See *Estimates Committee, 1977-78*, p. 222; and *Report of the Working Group on Energy Policy: 1979*, p. 65.

33. See Ramana, "Inevitability of Atomic Energy," pp. 221-38.

34. Ibid., p. 226. See also the section on the economics of nuclear power in the *Performance Budget of the Department of Atomic Energy, 1980-81*, New Delhi, pp. 3-6.

35. For an argument along these lines, see Onkar Marwah, "India's Nuclear and Space Programs: Intent and Policy," *International Security* 2, no. 2 (Fall 1977), pp. 106-13. See also Onkar Marwah, "The Economics of Nuclear Power," in James E. Katz and Onkar S. Marwah, eds., *Nuclear Power in Developing Countries* (Lexington, Mass.: Lexington Books, 1982), pp. 43-54.

36. For the rationales underlying the distribution of Indian defense funds, see Raju

G.C. Thomas, "The Armed Services and the Indian Defense Budget," *Asian Survey* 20, no. 3 (March 1980), pp. 280-97.

37. From Department of Atomic Energy, *Report: 1981-82.*

38. In November 1988, the price of Arab Gulf crude was $10.50 a barrel, and the price of American crude (West Texas intermediate) was $13.50 a barrel. See *New York Times*, Nov. 11, 1988.

39. From *India: 1986*, p. 456.

40. See Ministry of Energy, *Report: 1981-82*, p. 13.

41. *New York Times*, Nov. 4, 1988.

42. See Jones, *Nuclear Proliferation*, pp. 49-53.

3

South Korea

TONG WHAN PARK

Despite the global trend of declining interest in nuclear energy, South Korea continues to invest heavily in building new reactors to generate electricity. Although the first reactor was built as recently as 1978, South Korea is already a major producer of nuclear energy; measured by the actual level of generation, its nine operating reactors currently meet more than half of the domestic demand for electricity. By 1999, five more nuclear power plants will go into operation. When all fourteen reactors become fully operational, they will make South Korea one of the top ten countries in the world in the generation of nuclear-powered electricity.

South Korea's dependence on nuclear energy is the result of conscious planning by the government. It goes without saying that the onset of the oil crisis in 1973-74 served as a catalyst for Seoul's decision to go nuclear. For a country that had based its industrialization on imported petroleum, nuclear energy was probably the only viable alternative to hedge against future oil shocks. What is unique is that South Korea has followed through with its program for nuclear energy development, while most countries have not. In fact, South Korea is the only country in the world that is still expanding its nuclear power generation. Are Koreans not sensitive to the environmental and health hazards of nuclear energy? Or is Seoul's bureaucratic inertia too strong to be affected by the real and potential dangers of nuclear accidents, as shown in the Three Mile Island and the Chernobyl plants? The answers to these questions are neither simple nor straightforward. Making the matter even more complex, there still exists a lingering doubt about the possibility of South Korea's development of its own nuclear weapons.

Regardless of the undesirable side effects of nuclear power, South Korea's topmost priority is the security of energy supply to sustain its economic growth. Without energy supplies, the survival of the nation itself is threatened. Like any country that depends on imported energy, South Korea faces two critical challenges in solving its energy problems. One is the restructuring of its energy sources to reduce the heavy reliance on foreign oil. The other is to secure an uninterrupted supply of petroleum through diversification of suppliers and establishment of interdependent links with petroleum-exporting nations. Success in the first task hinges upon the development of alternative sources of energy, with special emphasis on nuclear-generated electricity. The second task would require a set of creative and carefully orchestrated policies toward oil exporters.

The purpose of this chapter is to analyze the political economy of energy in South Korea in the context of these policy goals. Specifically, this chapter examines: (1) the process of energy-driven industrialization in Korea; (2) the development of alternative sources of energy to replace and/ or complement petroleum; (3) the issues of external and internal security in light of various sources of energy; and (4) the prospects and problems of Korea's potential for nuclear weapons development.

The 'Miracle of the Han River'

After his rise to power in 1961, the late President Park Chung-Hee began preaching the doctrine of "nation-building through industrialization and exportation." He saw in industrial exports the only path for South Korea's survival in the second half of the twentieth century. To him, economic development was not merely a legitimizing device for his regime. Instead, it reached the status of a religion. Without Park's charisma and genuine belief in the doctrine, South Korea might not have become a leader among the Newly Industrializing Countries (NICs) in such a short time span.

It is widely held that during the first three Five-Year Economic Development Plans (1962-76), South Korea made the transition to a modern industrial state. Often called the miracle of the Han River, the country's performance in export-led industrial expansion during this period was phenomenal indeed. A World Bank report on South Korea described its industrial growth in these words:

> Manufactured exports rose rapidly in the early 1960s, albeit from a small base, but the real "turning point" in both export and industrial growth came around 1965, during a period of trade liberalization and other major policy reforms. In the decade following 1965, manufactured export growth coupled with rising domestic demand fueled industrialization much faster than before. The compound annual rate of growth in the index of manufacturing output was 11% from 1955 to 1965; it increased to 24% from 1965 to 1975. Underlying the acceleration of manufacturing output growth, the share of exports in manufacturing (gross) output, which was nil in 1955, rose from roughly 6% in 1965 to nearly 25% in 1975. Within a decade, from 1965 to 1975, the ratio of total exports to GNP more than trebled and the share of GNP originating in the manufacturing sector more than doubled. Manufactured products constituted 42% of total exports in 1965 and 74% in 1975.[1]

Clearly, South Korea's economy owed its takeoff to the structural change in patterns of production. Mining and manufacturing sectors (construction industry excluded) increased their share from 11.7 percent of the GNP in 1962 to 33 percent in 1977, while the agriculture, forestry, and fisheries sectors saw a sharp drop in their aggregate share, from 43.6 percent in 1962 to 21.7 percent in 1977. In particular, the heavy and petrochemical industries led the modernization drive in South Korea's economy. The output of iron and steel products increased 33-fold from 1961 to 1973. The machinery industry also achieved vigorous growth by adopting the tactics of mass production, specialization, and vertical integration. The result was sharply increased output during the period 1961-72: growth of 2.3 times in general machinery, 14.6 times in electrical machinery, and 9.8 times in the transport industry. Production capacity of the chemical fertilizer industry rose more than 17 times from 1961 to 1973. The petrochemical industry, established in 1968, paralleled the development of the oil industry.

The economic growth and the rapid progress in industrialization created higher demand for electric power, and the Seoul government responded by encouraging public and private investment in electric power projects. During the fifteen-year period of the first three five-year plans (1962-76), twenty-nine electric power plants were added to the existing roster of thirteen—seven hydroelectric and six thermal. The additional generating capacity was mostly thermal: twenty-three new thermal plants, compared with only six hydroelectric plants. In 1977, power generation capacity totaled 5.79 million kilowatts (kw), a 15.8-fold increase from 1961. As shown in figure 1, hydroelectric capacity declined from 33.1 percent of the total in 1961 to 12.3 percent in 1977, while the thermal power share rose from 66.9 percent to 87.7 percent. (See figure 3.1.)

During the early stage of building new thermal power plants, coal was to be the primary fuel. In the late 1960s, however, coal began to be replaced by petroleum. While in 1962 coal accounted for three-fourths of the total fuel for South Korean power plants, petroleum accounted for more than 91 percent of total energy supply in 1977.

Paralleling the rising use of petroleum in electricity generation, the overall level of energy consumption increased exponentially and its pattern changed drastically during the period 1962-76. Total energy consumption in South Korea climbed from 10.3 million MT of oil equivalent to 30.3 million MT, almost a threefold jump in fifteen years. The pattern of energy consumption changed mainly in the area of petroleum (figure 2). During the fifteen-year period from 1962 to 1976, oil consumption increased phenomenally, from 1 million MT to 17.8 million MT. On the other hand, consumption of firewood dropped from 5.3 million MT of oil equivalent in

Figure 3.1. Electric power generating capacity in South Korea
1961 - 1977

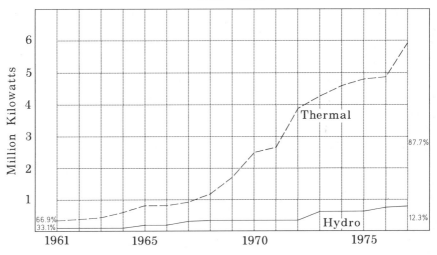

Source: Korea Electric Power Co.

1962 to 3.2 million MT in 1976, a 40 percent decrease. Coal and hydro-electric power consumption rose 2.3 and 2.5 times, respectively, during the same period. Consequently, the shares of the four primary energy sources in South Korea changed radically from 1962 to 1976. Petroleum (including natural gas) increased its share of total energy consumption from 9.8 to 58.8 percent, while firewood fell from 51.7 to 10.5 percent, coal from 36.8 to 29.3 percent, and hydroelectric power from 1.7 to 1.5 percent. Essentially the same pattern held true from 1976 to 1983, with only minor fluctuations from year to year. The respective shares in 1983 were 56.2 percent for petroleum, 33.1 percent for coal, 1.4 percent for hydro-electricity, 4.8 percent for firewood and charcoal, and 4.5 percent for nuclear-generated electric power. During the same period, total energy consumption rose by 164 percent. After 1983, there was a marked shift away from dependence on oil in favor of nuclear energy. By 1986, pe-troleum's share of total energy use fell to 46.7 percent, while that of nuclear energy rose to 11.6 percent. Hydroelectricity was little changed, at 1.7 percent, while coal and firewood accounted for 37.6 and 2.4 percent, respectively. South Korea started using liquified natural gas (LNG) in 1986, and its share increased from 1.0 percent to 3.0 percent within a year.[2]

In an attempt to satisfy the increasing demand for petroleum products, South Korea built its first oil refinery in 1964, in equal partnership with the Gulf Oil Corporation of the United States. Known first as Yukong Limited and later as the Korea Oil Corporation, this joint venture started commer-

Figure 3.2. South Korean consumption of primary energy

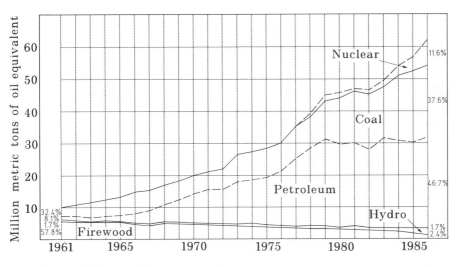

Source: Ministry of Energy and Resources

cial operation at the Ulsan refinery at a capacity of 35,000 barrels per day (bpd). Since the economic development plans required a stable supply of petroleum products, Korea proceeded to construct four more refineries: Kukdong Oil Co., Ltd. went on stream in 1966; Honam Oil Refinery Co., Ltd. in 1969; Kyungin Energy Co., Ltd. in 1972; and Ssangyong Oil Refining Co., Ltd. in 1980. As shown in table 1, Korea's petroleum refining capacity has grown in leaps and bounds to 790,000 bpd.

With the rising demand for oil as the major source of energy for industrial and public use, the volume of imported crude oil increased steeply: from 5.8 million barrels in 1964 to over 69 million barrels in 1970, then to almost 183 million in 1981. Like most importing countries, South Korea reduced its crude oil imports after the oil crisis of 1979-80, to a level of 178.4 million barrels in 1982. With the phenomenal economic growth in the 1980s, however, South Korea had to increase its oil imports thereafter—from 193 million barrels in 1983 to 230 million in 1986. The total dependence on foreign suppliers for crude oil has imposed a heavy burden on South Korea's foreign exchange reserve.

Sudden upsurges in petroleum prices following the two international oil crises caused foreign exchange spent for crude oil imports to jump from $305 million in 1973 to $2 billion in 1977 and then to over $6 billion in 1982. (See table 2.) In 1982, crude oil imports amounted to 25.1 percent of the total import bill and equaled 27.8 percent of South Korea's export revenues. During the period 1964-77, the most critical era in South Korea's

Table 3.1. Petroleum refining capacity in South Korea
            (Thousands of barrels per day)

	Total	Yukong	Honam	Kyungin	Ssangyong	Kukdong
1964	35	35	—	—	—	—
1965	35	35	—	—	—	—
1966	40	35	—	—	—	5
1967	60	55	—	—	—	5
1968	120	115	—	—	—	5
1969	180	115	60	—	—	5
1970	220	115	100	—	—	5
1971	270	115	100	50	—	5
1972	400	175	160	60	—	5
1973	400	175	160	60	—	5
1974	440	215	160	60	—	5
1975	440	215	160	60	—	5
1976	440	215	160	60	—	5
1977	440	215	160	60	—	5
1978	580	280	230	60	—	10
1979	580	280	230	60	—	10
1980	640	280	230	60	60	10
1981	790	280	380	60	60	10
1982	790	280	380	60	60	10
1983	790	280	380	60	60	10
1984	790	280	380	60	60	10
1985	790	280	380	60	60	10
1986	790	280	380	60	60	10

Source: Ministry of Energy and Resources.

economic development, annual imports of petroleum increased 26.5 times, from 5.8 million to 154.5 million barrels. In the meantime, outlays for foreign crude increased 160 times, from $12.5 million to $2 billion.

Uncertainties about international oil prices intensified during this period, when petroleum exporters wielded the "oil weapon" as an instrument of foreign policy. It was fortunate for South Korea that the price of crude oil started to decrease in 1983 as part of the global phenomena known as the "three lows"—the other two areas of decline being interest rates and the value of the U.S. dollar. The cost of oil in 1986 for Seoul was less than $3.5 billion despite the import volume of over 230 million barrels.

South Korea is handicapped in oil procurement by its limited number of sources. As shown in table 3, Kuwait supplied 100 percent of South Korea's petroleum needs in 1964, and until 1968 its sole suppliers were the two Middle Eastern countries of Kuwait and Iran. Saudi Arabia began shipping crude oil to South Korea in 1969 and was a dominant source from 1973

Table 3.2.   South Korea's bill for crude oil imports

	Import volume (000 bbl)	FOB basis		C & F basis	
		Total (000)	$/bbl	Total (000)	$/bbl
1964	5,835	$ 9,426	$ 1.62	$ 12,607	$ 2.14
1965	11,170	18,092	1.62	23,677	2.12
1966	14,202	22,810	1.61	29,803	2.10
1967	18,638	28,852	1.55	37,909	2.03
1968	36,409	52,637	1.45	70,683	1.94
1969	55,889	77,621	1.39	97,681	1.75
1970	69,150	95,918	1.39	118,916	1.72
1971	85,425	147,944	1.73	178,761	2.09
1972	92,581	176,320	1.90	221,077	2.39
1973	103,210	253,020	2.45	305,158	2.96
1974	112,703	1,013,522	8.99	1,104,833	2.80
1975	117,795	1,241,214	10.54	1,328,152	11.28
1976	132,407	1,492,011	11.27	1,580,003	11.93
1977	154,549	1,890,104	12.23	2,000,075	12.94
1978	166,532	2,042,764	12.27	2,173,680	13.05
1979	185,153	3,153,352	17.00	3,330,608	17.95
1980	182,861	5,431,438	29.70	5,641,434	30.85
1981	180,316	6,237,384	34.12	6,504,165	35.58
1982	178,369	5,854,300	32.82	6,075,324	34.06
1983	192,888	5,550,319	28.77	5,767,996	29.90
1984	199,681	5,587,105	27.98	5,807,394	29.08
1985	198,313	5,289,381	26.67	5,499,593	27.73
1986	230,063	3,259,118	14.17	3,474,588	15.10

Source: Ministry of Energy and Resources.

to 1984, when Iran became the largest supplier. After 1983, the Seoul government accelerated its efforts to diversify the sources of oil supply, reducing the dependence on the Middle Eastern oil from 74 percent in 1983 to 60.5 percent in 1986.

The strong association between South Korea's industrialization process and its rising dependence on crude oil is clear. Not only did the GNP grow along with petroleum consumption, but the rate of increase in manufacturing output closely matched that of crude oil imports. During the period 1965-75, for example, the annual rate of growth in the compound index of manufacturing output was 24 percent, while the average annual growth rate for oil imports was 28.9 percent. According to the forecasts of the Seoul government, South Korea's dependence on foreign oil will remain high through the 1990s, close to the 1986 level of 47 percent.[3] There are three basic reasons for this outlook. First of all, domestic coal production has reached a plateau, due to the combination of a poor resource base and

Table 3.3.   Sources of South Korea's crude oil imports
            (Thousands of barrels)

	Middle East					Asia	Africa	Middle & South America	Total
	Saudi Arabia	Kuwait	Iran	Oman	Other				
1964	—	5,835	—	—	—	—	—	—	5,835
1965	—	7,137	4,033	—	—	—	—	—	11,170
1966	—	8,513	5,689	—	—	—	—	—	14,202
1967	—	11,209	7,435	—	—	—	—	—	18,638
1968	—	17,997	18,412	—	—	—	—	—	36,409
1969	12,545	24,464	18,880	—	—	—	—	—	55,889
1970	21,995	24,998	22,197	—	—	—	—	—	69,150
1971	30,548	43,984	10,802	—	—	—	—	—	85,425
1972	30,867	46,628	2,824	—	6,262	—	—	—	92,581
1973	63,411	19,401	4,554	—	15,844	—	—	—	103,210
1974	73,713	18,866	3,305	—	16,799	—	—	—	112,703
1975	51,867	53,637	882	—	11,415	—	—	—	117,795
1976	52,640	52,406	14,234	—	13,127	—	—	—	132,407
1977	83,396	49,657	16,766	—	4,730	—	—	—	154,549
1978	95,840	50,789	12,912	—	6,991	—	—	—	166,532
1979	98,865	59,887	19,809	—	6,953	154	—	—	185,513
1980	111,853	47,612	15,515	—	7,881	—	—	—	182,861
1981	104,900	38,800	14,500	—	7,700	5,100	1,500	10,300	182,800
1982	88,500	21,300	22,900	—	2,800	21,900	6,400	14,500	178,400
1983	58,900	20,700	31,400	17,900	14,000	26,300	12,100	11,700	193,000
1984	35,800	15,800	39,800	24,100	17,200	38,500	12,000	16,200	199,700
1985	13,700	17,500	34,100	23,100	24,500	48,000	11,900	25,400	198,300
1986	21,300	13,200	37,400	27,800	39,600	51,200	14,000	25,600	230,100

Source: Ministry of Energy and Resources.

unfavorable mining conditions. Rising income will also lead to heating of living quarters with oil rather than anthracite coal. Second, South Korea will experience an exponential growth in the transportation industry, especially for privately owned passenger cars. The third factor is South Korea's limited potential in alternative sources of energy. The development potential for hydroelectric and tidal power is extremely small. Plagued with technical and economic problems, solar energy will not make a significant contribution to South Korea's energy supply in the foreseeable future. Except for nuclear energy, it appears that Seoul has little latitude in the development of alternative energy sources.

NUCLEAR ENERGY AS A VIABLE ALTERNATIVE

Korea's GNP is expected to grow at an annual average rate of 7 percent through the end of this century, according to a recent forecast made by the

Table 3.4.   Long-term energy demand forecast for South Korea
(Provisional; thousand tons of oil equivalent)

	1986	1991	2001	2010
Petroleum	28,500 (46.7%)	38,100 (47%)	57,900 (47%)	70,800 (43%)
Anthracite Coal	12,800 (21.0%)	12,800 (16%)	7,500 (6%)	5,100 (3%)
Bituminous Coal	10,100 (16.5%)	13,300 (16%)	33,700 (27%)	51,500 (31%)
Nuclear	7,100 (11.6%)	12,500 (15%)	18,400 (15%)	27,300 (16%)
LNG	100 (0.2%)	2,600 (3%)	2,600 (2%)	6,500 (4%)
Hydro and firewood	2,500 (4.0%)	2,700 (3%)	3,200 (3%)	4,800 (3%)
Total	61,100 (100%)	82,000 (100%)	123,400 (100%)	166,000 (100%)

Source: Ministry of Energy & Resources.

government. Energy demand will rise by approximately 5 percent annually until the year 2001, then by only 3.4 percent from 2002 to 2010. The reliance on imported energy will likewise increase, from 66.5 percent in 1986 to 76.4 percent in 2001 and 77.6 percent by 2010.[4]

Confronted with growing dependence on imported energy, South Korea has been seeking a structural transition in the composition of energy sources. The main thrust has been to reduce the shares of petroleum and anthracite coal while increasing those of nuclear energy, bituminous coal, and LNG. Nuclear energy accounted for only 2 percent of the country's energy demand in 1982 but rose to 11.6 percent in 1986. In September 1989 the ninth nuclear reactor went into operation, bringing installed nuclear capacity to 7.62 million kilowatts (kw), representing 36.5 percent of Korea's total electricity generating capacity. By 2001, nuclear power will meet a projected 15 percent of total energy needs. Bituminous coal, which raised its share from 12.2 percent of total energy demand in 1982 to 16.5 percent in 1986, will reach 27 percent in 2001, according to provisional energy forecasts (table 4). On the other hand, the Seoul government's energy policy calls for a declining share of the energy mix for petroleum. Whereas oil accounted for 57.8 percent of energy used in 1982, its contribution was lowered to 46.7 percent in 1986 and will remain at that level throughout the 1990s.

To achieve such an ambitious transition in energy configuration, Seoul's planners employed a number of strategies, including the diversification of oil supply sources, "development import" of oil and coal from overseas, and a concentrated investment in nuclear energy. The centerpiece of Seoul's plan is the controversial program to increase its dependence on nuclear-generated electricity. This approach seems to have been working so far, even though it has seen its share of problems.

For a rapidly industrializing country like South Korea, the supply of electricity for industrial use is the backbone of sustained economic growth. Confronted with the reality of an unstable oil market, Koreans have accepted nuclear power as the primary alternative for the supply of electricity up to the year 2000. In 1981, nuclear reactors accounted for only 6 percent of the electricity generated, while oil-powered plants supplied 74.2 percent of demand. By 2001, Seoul hopes to lower oil's share of power generation to a meager 11.3 percent, while raising nuclear power to 36.1 percent of total supply and keeping it there. The relative proportion of hydroelectric plants will remain stable, while coal-fired power generation is expected to rise.

Korea's decision-makers saw three opportunities in this approach. First, relief from the burden of heavy reliance on foreign petroleum is desirable from the political-economic standpoint. The plan is especially attractive because it complements nicely Korea's attempt to explore its own sources of crude oil. The second opportunity is the internalization of nuclear technology. No major breakthrough has yet been made in the search for alternative sources of oil that are both safe and inexpensive. Solar, geothermal, and other exotic energies will make only limited contributions during the next decade. Given the premise that nuclear power is perhaps the only viable alternative to petroleum in the foreseeable future, it becomes imperative for the South Koreans to internalize the nuclear technology. The third opportunity in "going nuclear" takes advantage of the severely shrinking demand in the international nuclear industry. For example, the industry received orders for only two reactors in 1978—both from South Korea, while the global nuclear industry has the total capacity for building 15 to 20 generators per year. In such a buyer's market, South Korea can bargain for the best possible deal by playing one supplier against the others.

There are some serious problems, however, in an energy development plan that puts an excessive weight on nuclear power. Nuclear reactor technology is controlled by a small number of suppliers, upon which South Korea will have to remain dependent for a considerable period of time. Only a handful of foreign companies are involved in the nuclear power plants under construction or in operation in South Korea (table 5). Since Koreans lack the know-how for nuclear power generation, there is no assurance that they will be treated fairly by these foreign firms. A highly publicized scandal over claims of overcharging by the Bechtel Corporation is eloquent testimony to Seoul's vulnerability. According to the heated debate in the National Assembly's Trade and Industry Committee, the Korea Electric Power Corp. (KEPCO) concluded "unfavorable contracts" with the Bechtel Group for the design and construction of four nuclear power plants in South Korea. Opposition lawmakers called for the revision

Table 3.5.   Nuclear power plants in South Korea
              (February 1989)

Plant	Capacity (kw)	Type	Startup
KoRi 1	556,000	PWR	1978
KoRi 2	605,000	PWR	1983
KoRi 3	895,000	PWR	1985
KoRi 4	895,000	PWR	1986
Wolsong	629,000	PHWR	1983
Yongkwang 1	900,000	PWR	1987
Yongkwang 2	900,000	PWR	1988
Ulchin 1	920,000	PWR	1988
Ulchin 2	920,000	PWR	1989
Yongkwang 3	950,000	PWR	Under construction
Yongkwang 4	950,000	PWR	Under construction

Source: *Nuclear News*, February 1989, pp. 75-76.

of the bilateral contracts to avert future incidents of overcharging. In part due to this political pressure, KEPCO received in 1984 a refund of close to $5 million of the $16 million allegedly overpaid to Bechtel. At the risk of oversimplifying the complex case involving the Bechtel Corporation, the cause may be the practice by which "Bechtel submits bills and KEPCO pays."

A second drawback for nuclear power in South Korea is that the fuel must be imported, at the cost of large amounts of foreign exchange, since Korea does not produce uranium at an economically profitable level. The Ministry of Energy and Resources has estimated that uranium imports in 1987 totaled 298 tons at a cost of $192 million; in 1988, the total was 342 tons of uranium oxide costing $312 million.[5] Moreover, there is no guarantee that uranium-exporting countries will refrain from the pattern set by some oil-exporting nations in raising prices and threatening embargoes. As long as South Korea remains an ally of the United States, however, the danger of nuclear fuel shortage is not likely to materialize.

The third obstacle is the environmental hazard posed by nuclear reactors and wastes. South Korea has experienced no major reactor accidents to date, but that does not mean that the country's safety record has been above the international standard. In 1987, routine check-ups of the reactors in operation revealed fifty-five violations and inadequate observations of safety rules. As a result, there were twenty-six unscheduled shutdowns in South Korea in one year. The frequency of unscheduled shutdowns was 3.7 per reactor, comparable to that of the U.S.—3.9 per reactor in 1986. But the incidence in South Korea was seven times higher than Japan's 0.5 shutdowns per reactor.

The primary reason for South Korea's nuclear safety problem is organizational: the government body charged with safety inspection lacks institutional independence, as well as trained personnel. The highest responsibility for nuclear safety lies with the Atomic Energy Commission, which reports to the prime minister. Of the six members of the commission, four are ex officio: the minister of the Economic Planning Board, the minister of science and technology, the minister of energy and resources, and the president of KEPCO. The inclusion of the latter two casts doubt about the neutrality of the commission's work, for they are the key architects and managers of nuclear power plants. In a case in point, the commission made a fifty-two-point recommendation to upgrade reactor safety after the Three Mile Island accident in 1979, but eight recommended items have yet to be accepted by KEPCO on the grounds that the measures would cost over $12.5 million per reactor.

It is no accident that many South Koreans are advocating a change of status for the commission. Instead of operating under the office of the prime minister, they argue, the commission should report directly to the president, as does the Nuclear Regulatory Commission in the United States. The Achilles heel of South Korea's nuclear safety, however, is the control of the Atomic Energy Safety Center, charged with the task of inspecting nuclear power plants for environmental hazards. The center is affiliated with the Korea Advanced Energy Research Institute, which not only depends on KEPCO for funding but also has KEPCO's president as its chairman of the board. With such an institutional setup, it is difficult to conduct safety inspections impartially and independently.

Balancing its advantages with its problems, Korea's nuclear development plan appears to be a mixed blessing. From the viewpoint of energy supply, the plan seems quite adequate despite the weakness in safeguards. From the larger perspective of the global political economy, there is room for reconsideration. During the next two decades, OPEC's member nations will undergo the process of industrial revolution. Along with the capital acquired in the form of petrodollars, they will need technology and skilled labor. It is the latter requirement on which South Korea may continue to capitalize. By tying purchases of crude oil to the supply of technology, South Korea could enjoy a competitive edge in a scramble for the OPEC market. But an enhanced linkage with oil exporters need not lead to a higher dependence on imported oil, for South Korea's petroleum dependence is already too high.

ENERGY, SECURITY, AND THE POTENTIAL FOR NUCLEAR WEAPONS

In an analysis of South Korea's energy outlook and strategy for the future, it is imperative to examine the problem of security as it relates to energy:

specifically, the question of safeguarding energy sources and energy-related installations. South Korea faces threats from both without and within, and any damage inflicted upon its energy system will have far-reaching repercussions beyond the losses themselves. At stake will be not only the survival of a regime but also the viability of the state itself. An in-depth look at the sources of threat to South Korea's energy supply sheds light on the directions the government may take to maximize energy-related security. It also gives a glimpse of past attempts and the current posture of government planners toward development of an indigenous nuclear weapons capability.

Two major considerations govern Seoul's strategic thinking about the external security of energy. One is the safety of the sea lanes through which oil tankers pass, while the other is the constant threat of attack by North Korean infiltrators. As for the shipping routes, there is little South Korea can do but act prudently to minimize the risks. The country's dependence on Middle Eastern oil has been so great that even during times of crisis in the Persian Gulf, its imports from the region have continued at the normal rate. The government has emergency plans in the unlikely event that the Persian Gulf is closed, however. The Ministry of Energy and Resources predicts that its reserve oil stockpile would carry South Korea for 102 days and could be stretched to cover 150 days should the emergency plan go into effect. The stockpile contains 23.8 million barrels of crude oil and 12.4 million barrels of petroleum products, equivalent to 65 days' use at normal rate of consumption. In addition, South Korea has a standing commitment from the United States to help secure oil in case of emergencies.

Despite the sizable stockpile and the U.S. pledge of assistance, however, disruption of oil supply would have a highly negative impact on South Korea's economy. One report of the Korea Development Institute forecasts that a 20 percent drop in oil supply for six months would reduce national income by 11.9 percent, cause unemployment for 1.35 million people, and reduce exports and imports by 12.9 and 24.1 percent, respectively. Even more pessimistic, the government's think tank projects that if oil supply is reduced by 30 percent, the effects on national income and unemployment will be almost double those of a 20 percent reduction.[6]

Whereas the possible closure of the Persian Gulf poses a severe threat, its impact would not be limited to South Korea. In fact, such an eventuality would most likely bring about a global crisis over which Seoul could exercise little control. From Seoul's standpoint, no less menacing than a crisis in the Gulf is the threat from North Korea against energy supply and installations in the South. One of the major targets would be oil and LNG tankers en route to South Korea. Vessels owned or chartered by South Koreans are easy prey for surprise attacks by North Korean guerrillas, as witnessed in the November 1987 terrorist attack on Korean Air Flight 858.

It is widely known that Pyongyang is undergoing a critical transition of power from Kim Il-Sung to his son Kim Chong-Il. In an attempt to divert domestic tensions about the succession, Pyongyang may ignite confrontations with the South. It has a contingent of 100,000 specially trained commandos that can be deployed for any type of unconventional warfare. South Korea has uncovered four underground tunnels in the Demilitarized Zone that could have been used for the infiltration of troops, and it is estimated that North Korea is digging more than twenty such tunnels in the hope that some may evade detection. Moreover, Pyongyang has often sent guerrillas with underwater gear aboard speedboats, who could cause untold damage should they be successful in attacking the operating nuclear reactors in South Korea, which are located in or near the coastal areas, within easy reach of seaborne attackers. Beside nuclear reactors, oil refineries and conventional power plants could be crippled by North Korean commandos. Seoul has a strong military force and a well-trained militia, of course, and it places a high priority on the protection of its industrial sites. But the unfortunate reality is that South Korea has a relatively open society compared with that to the north. North Korea's "special troops," trained in the southern dialect and style of living, might be able to penetrate South Korea's defense and blow up energy stockpiles and processing plants.

An attack by Pyongyang on South Korea's energy-related installations would most probably invoke physical sanctions from Seoul. The tricky point is that North Korea would camouflage its actions to make it difficult to identify the culprits. Even after the 1983 massacre of Rangoon, in which seventeen top-ranking South Korean government officials were murdered, Pyongyang vehemently denied its involvement, despite the fact that its commandos were captured and convicted of the crime by the Burmese court. The international politics of Northeast Asia will also constrain Seoul from launching a massive retaliatory attack against the North for fear of destroying the quadrilateral balance of power. After all, nobody can deny that a regional status quo has been pursued by the United States, the Soviet Union, China, and Japan.

Put these considerations into the decision matrix of the Seoul government and it is understandable why Seoul is sensitive to the role of U.S. troops stationed in the country. Without U.S. forces serving as a tripwire against a North Korean invasion, South Koreans face a justifiable "window of vulnerability" vis-à-vis the Northerners. It is an irony that Seoul's economic miracle hangs on a thin line of military security sustained in large part by some 43,000 U.S. troops.

There is no denying that South Korea has had domestic problems ranging from the skewed distribution of wealth to the oppression of political dissidents; mismanagement of energy and other critical resources may have contributed to domestic instability. Despite internal problems, however, South Korea has not seen a major threat to its energy-related installations. No known cases of sabotage have taken place in power plants, nuclear or other. The most serious case of unrest occurred in a labor dispute at the Sa-Buk coal mine in 1980, which was handled with dispatch by the government. Even when labor disputes swept the country in 1987, none of the energy-producing facilities were affected seriously. It is prudent to say that South Koreans will not direct their discontent at energy-related facilities because energy represents their lifeline. This does not imply, however, that the government's energy policy is immune to popular scrutiny. If anything, the Seoul government has opened itself to public opinion by wisely subjecting its energy policies to public debate. Not only does the government want to see a consensus emerge on energy issues, but also it seeks to share the responsibility of energy shocks with the larger segment of the society. Energy is such an issue of life and death, yet so difficult to control, that no government could shoulder the burden alone.

A question that is very much a part of South Korea's energy and security outlook is whether it will develop its own nuclear weapons capability. Possession of nuclear arms might arguably enhance Seoul's security posture. It is one thing to rely on the American nuclear umbrella but South Korea would acquire an entirely different status if it became a member of the prestigious nuclear club. There has been speculation about Seoul's posture toward the development of nuclear weapons. During the late President Park Chung-Hee's reign, public debate on the issue was tolerated if not encouraged. The Fifth Republic under President Chun Doo-Hwan saw no public comment on South Korea's becoming a nuclear power; the government's official posture was to support the Non-Proliferation Treaty of 1968, which was ratified by the Korean National Assembly in 1975. Nevertheless, it is assumed by many observers of South Korean affairs that Seoul could build nuclear devices without much difficulty should it so desire. The actual implementation of weapons production, however, would require a major change in the international climate of Northeast Asia.

It was the deteriorating security environment of Northeast Asia that prompted the Park regime to explore the nuclear option in the first place. Shocked by the Nixon Doctrine of 1969 and the decision to withdraw the Seventh U.S. Infantry Division from Korea, Park's government had little choice but to launch a program of greater self-sufficiency in national defense. In 1970, the government established two organs for this purpose: the Agency for Defense Development (ADD), charged with modernizing the weapons system, and the Weapons Exploitation Committee, to obtain high-powered weapons including atomic devices. With the fall of Vietnam, the credibility of America's extended deterrence declined even further.

At about the same time, North Korean President Kim Il-Sung visited Peking on April 15, 1975, amid speculation that he might be seeking China's approval for another invasion of the South. Confronted with such an unstable situation, President Park made his first statement on June 12, 1975, about the possible development of nuclear weapons in the case of the withdrawal of America's nuclear umbrella. He reaffirmed his statement on nuclear weapons in an interview with the *Washington Post* (June 25, 1975), while the Minister of Science and Technology, Choi Hyung-Sup, gave public support to President Park's remarks.

There are two critical components that a "threshold" country like South Korea must acquire before it can become a nuclear power: the nuclear device itself and the delivery system. The latter posed no barrier to potential nuclear proliferation by the Park government, which had F-4D/E fighter-bombers capable of carrying nuclear warheads. In 1978, the ADD successfully test-fired the first ground-to-ground guided missiles, with an estimated range of 100 miles.[7] Given the capability of delivery, it was in the nuclear fuel cycle that the Park government concentrated its efforts.

Though South Korea is known to have deposits of natural uranium, the quantity is not significant. According to a 1981 report of the Korea Institute of Energy and Resources (KIER), U_3O_8 reserves were estimated at 10,000 metric tons—barely enough to fuel three nuclear power plants for 30 years.[8] It was only logical that South Korea seek to buy a reprocessing plant for spent nuclear fuel. In fact, Seoul's intention to obtain reprocessing capability dates back to 1968, when such a project was first set up as part of the Long-term Plan of Research, Development, and Use of Nuclear Energy (1968-83). In 1972, the Korea Atomic Energy Research Institute[9] drew up a detailed plan to construct a reprocessing plant by 1977. In May 1972, negotiations began between the South Korean and French governments on nuclear fuel fabrication and a reprocessing plant, as well as cooperation in nuclear research. Talks continued at the working level and resulted in the bilateral Agreement for Technical Cooperation in Atomic Energy, which became effective on October 19, 1974.

While the deal was being cut with the French government, Seoul ratified the Treaty on the Non-Proliferation of Nuclear Weapons on March 20, 1975. The ratification was intended to convey to the United States and to other regional powers that South Korea would not develop its own nuclear weapons. In a sense, it was a political ploy that would help allay fears of the American government, which was deeply concerned with the ongoing talks between Seoul and Paris. Korea and France also signed,

under the International Atomic Energy Agency (IAEA), the Agreement
for the Application of Safeguards, to be effective from September 22, 1975.
Meanwhile, the French government notified Seoul of its readiness to
provide $20 million for nuclear reprocessing facilities, and Korea was
about to acquire its own reprocessing program.

The U.S. government had decided, however, that the French-Korean
deal was not to materialize. Starting in November 1975, the Americans
initiated a series of discussions with Korean officials about the reprocessing
plant project. Washington's pressure tactic worked and on January 29,
1976, Myron B. Kratzer, U.S. acting assistant secretary of state for oceans
and international environmental and scientific affairs, disclosed at a Senate
Government Operations Committee meeting that South Korea had canceled its plans to purchase a French plutonium reprocessing plant. In an
attempt to save face, French President Giscard d'Estaing said on NBC's
"Meet the Press" on May 23 that he had decided against selling the
reprocessing plant before South Korea canceled its purchase.[10]

It has been reported that Seoul has continued to show interest in
obtaining reprocessing technology. In 1984, a proposal was allegedly made
by the Canadian Atomic Energy Agency to recycle the spent fuel from a
U.S.-made light water reactor (LWR) in South Korea into mixed oxide fuel
(MOX), which would contain weapons-grade plutonium for the Canadian
heavy water reactor (HWR) at Wolsung.[11] The proposal was reportedly
blocked by the U.S. government. Due in part perhaps to this alleged
incident, the Seoul government allowed installation of an added security
device on the Wolsung CANDU nuclear power plant to account for the
disposition of all plutonium, since it requires only about 8 kilograms of
plutonium to make a nuclear bomb. Called the spent fuel sealing system,
this device serves as a backup to surveillance systems previously installed
in the spent fuel storage bays. The new system would provide the IAEA with
an additional guarantee that no diversion of plutonium would take place.

The Sixth Republic, launched on February 25, 1988, will continue to
honor the Non-Proliferation Treaty. Deeply committed to rapprochement
with China and the Soviet Union, President Roh Tae-Woo is not likely to
become a second Park Chung-Hee. Roh has also made it clear that he will
pursue the unification of Korea through peaceful means. Though he will
try to negotiate with the North Koreans from a position of strength, he is
not likely to authorize a nuclear weapons development program. At present, Seoul is in an optimum position to improve relations with its northern
neighbors, and the country is becoming an important trading partner for
the Communist bloc. On the other side, the United States is not expected
to withdraw its nuclear umbrella from Northeast Asia in the foreseeable
future and definitely does not want to see a "Korean bomb."

The challenges of energy management are multidimensional, complex, and dynamic. They are multidimensional because their causes and consequences affect virtually all aspects of a modern polity, whether economic, political, military, or sociocultural. They are complex, as the ingredients of energy policies interact with each other in a manner not amenable to simple linear analyses; energy-related issues are dynamic because they are in a state of constant change. For these reasons, programs of most countries are at best a compromise between long-term tactics and short-term trial and error. South Korea cannot be an exception to this rule.

Through the 1990s, Seoul's energy programs will be guided by four major goals. First, South Korea will continue to build a basis for secure supply of oil, meanwhile enlarging its stockpile for commercial and strategic uses. Second, Seoul's policies will aim at reducing its relatively high dependence on oil. Failing that, it will try to maintain oil's share in total energy consumption at a level below 50 percent. Growth in South Korea's energy consumption, in both residential and industrial sectors, will be met by such non-oil sources as nuclear power, coal, and LNG. Though modest in scale, research and development will continue on solar, wind, and tidal energy sources. Third, South Korea will accelerate its efforts to become a producer of energy, with oil and gas development vigorous on its continental shelf, in addition to overseas oil exploration. Seoul will continue its overseas investment not only to secure oil, but also to mine coal and other mineral resources. Fourth, despite the quantum rise in the living standard of its people, the government will not slacken its drive for energy conservation through a combination of incentives and sanctions.

The backbone of Korea's energy development through the end of this century will be nuclear-generated electricity. Such commitments have already been made and there will be no turning back in the foreseeable future, although there will surely be much controversy about safety and weapons development potential. The success of Seoul's nuclear-based energy policy will depend on the government's will to mix flexibility with firmness. It must maintain consistency in energy planning while remaining strong and credible enough to make swift adjustments to the changing environment. As experienced in the energy crises of the past, South Korea is but a small player in the volatile field of international energy. Its survival will depend upon skillful maneuvering in the political economy of the global energy market.

NOTES

The research reported in this chapter was supported in part by a grant from the Korean Traders Scholarship Foundation. The author wishes to express thanks to Ms. Young Sook Moon, who helped update the data for the period 1982-86.

1. Westphal, Larry E., "The Republic of Korea's Experience with Export-led Industrial Development," *World Development* 6, no. 3, (1978), pp. 347-82.

2. For a more detailed analysis, see Korea Energy Economics Institute, *Korea's Energy Future: The Long-term Perspective and Strategies* (1987-2010), in Korean (Seoul: Korea Energy Economics Institute, 1987), p. 71.

3. The Ministry of Energy and Resources has made various estimates, all of which converge on a continued dependence on oil. Only after 2001 does it envision even a slight drop in petroleum dependence, to about 42 percent of total energy.

4. This forecast is based on Korea Energy Economics Institute, *Korea's Energy Future*, p. 11.

5. Ministry of Energy and Resources, "The Energy Supply and Demand Forecast for 1988," *The Korea Petroleum Association Bulletin*, (February 1988), pp. 15-27.

6. Reported in *Choong-Ang Ilbo*, May 24, 1984, Chicago edition.

7. *Korea Herald*, Sept. 27, 1987.

8. *Korea Herald*, Feb. 25, 1981.

9. In January 1981, the Korea Atomic Energy Research Institute and the Korea Nuclear Fuel Development Institute were combined into the Korea Advanced Energy Research Institute.

10. For a more detailed discussion of Korea's failed attempt to buy a reprocessing plant, see Young-Sun Ha, *Nuclear Proliferation, World Order and Korea* (Seoul: Seoul National University Press, 1983).

11. Taylor, P., "Ottawa Denies U.S. Killed A-Deal," *Globe and Mail* (Toronto), Oct. 16, 1984, p. 5.

4

Pakistan

CHARLES K. EBINGER

Since its founding in 1947, access to vital energy supplies has played a major role in Pakistan's economic and its foreign policy. At independence, Pakistan had almost no industries, and its only power stations were three hydroelectric plants, two in the North-West Frontier Province (NWFP) and one in the Punjab.[1] During 1947-58, Islamabad acquired only 46 megawatts (Mw) of additional generating capacity, owing to the high costs of dam construction and its conflicts with India and Afghanistan over water resource rights.

With the implementation of the Indus Water Treaty of 1960, however, Pakistan obtained rights to the waters of the Indus, Jhelum, Kabul, and Chenab rivers. With international assistance, Pakistan embarked on the construction of two giant earthrock dams at Mangla and Tarbela and the development of extensive barrage and canal systems throughout the Indus Basin. Although Pakistan has a large hydroelectric potential, with estimates varying between 10,000 and 12,000 Mw, the expansion of hydel power in Pakistan is plagued by: (1) the large seasonal fluctuations (60-70 percent) in the flow of the rivers; (2) the impact that drought can have on hydroelectric generating capacity, intensified as irrigation receives priority over generation; (3) the high siltation rates of the Indus and its major tributaries, which limit the life of large-scale hydro plants; and (4) the high cost of dam construction and transmission and distribution facilities in the remote north, where hydel potential is greatest, far from the major load centers.[2]

Given these problems, Pakistan has developed a substantial thermal electricity capacity to offset shortfalls in hydel generation; thermal generation also provides electricity to areas of the country not served by hydel capacity. Nonetheless, only about 22 percent of the country's population has access to the electricity grid. Although natural gas reserves are substantial, at 530 billion cubic meters (BCM), and gas has received priority since the 1960s as a fuel for thermal electricity generation, serious shortages of natural gas occurred in 1983-84,[3] the result of a combination of subsidized prices to consumers, low producer prices at the wellhead, and skyrocketing demand. Even though Pakistan has significant coal reserves, nearly all are poor-quality lignite or subbituminous coal with limited coking potential, a low calorific content, and a high sulfur component. Moreover, because the coal is subject to spontaneous combustion when stacked and deteriorates upon exposure to air, it has until very recently[4] not been much in demand, becoming progressively less important in the national energy balance since 1965.

The problems plaguing the hydel, gas, and coal sectors and the rising prices for oil led to interest in nuclear power in Pakistan even before the 1973-74 OPEC price rises, as evidenced by Pakistan's acquisition from Canada in 1971 of its first nuclear power plant—a 137-Mw heavy water reactor outside Karachi. (In February 1990, France agreed to sell Pakistan its second nuclear power plant.)

ENERGY AND SECURITY: THE 1973-74 OPEC CRISIS

Still reeling under the economic dislocations engendered by the 1971 secession of East Pakistan, now Bangladesh, and a series of disastrous nationalizations in the first half of 1972, Pakistan was ill-equipped to handle the devastating economic impact of 1973-74 OPEC oil price rises.

The increases in the price of oil had both direct and indirect effects on the Pakistani economy. With crude oil and petroleum products in 1972 accounting for about 42 percent of the nation's commercial energy use, clearly Pakistan was vulnerable to oil price rises, as evidenced by the escalation in its oil import bill—from $50 million in 1973 to $540 million in 1979, $1.1 billion in 1979-80 following the Iran crisis, and $1.3 billion in 1982-83. Simultaneously, Pakistan's current account deficit rose from $131 million in 1972-73 to a peak of $1,187,000,000 in 1974-75 before tapering off to $538 million in 1982-83. However, in 1986 it escalated to $2 billion.[5]

Behind the drastic slide in Pakistan's current account position were: (1) the loss of income from East Pakistan's jute sales, (2) the escalation in the price of oil, (3) the enactment of ill-judged domestic economic policies, and (4) fundamental structural imbalances that continue to plague the Pakistani economy—a decrease in international prices for Pakistan's exports and stagnating domestic production, especially of cotton and textile goods. The Pakistani economy remains dependent on export earnings from rice, cotton and its products, and wheat. Because these crops are at the mercy of weather conditions, even relatively mild weather fluctuations can spell disaster. When the monsoon comes too early or too late, or floods the country, the national economy is in crisis.

Even prior to 1974, Islamabad had recognized the need to enact a vigorous oil exploration program. With domestic oil production filling only 13 percent of the nation's requirements, Pakistan since 1972-73 has promoted joint ventures in petroleum exploration with foreign oil companies, in close association with the state-owned Oil and Gas Development Corporation. To encourage participation, Pakistan has drawn up a model concession agreement and has moved steadily toward indexing the price of crude oil and petroleum products to international prices.[6] Likewise since 1973, Pakistan has moved to develop an oil products pipeline from the remote northern oil-producing and refining centers to Karachi and Hyderabad to replace imported petroleum. Despite these prodigious efforts, however, domestic oil production in 1982-83 accounted for only 11 percent of total consumption.[7]

By the end of the Fifth Five Year Plan (1978-83), Pakistan was self-sufficient to the extent of about 67 percent of its total energy supplies; the remaining needs were met exclusively by imports of crude oil and petroleum products. This degree of self-sufficiency has occurred owing to the steadily growing share of natural gas and hydel in the total energy supply mix.[8] The contribution of hydel and natural gas to the energy picture increased from 12.8% and 35.6%, respectively, in 1971-72 to 16.1% and 40.8% in 1981-82. In the same period, the respective shares of oil and coal fell from 42.9% and 8.3%, respectively, to 36.9% and 5.6%. Nuclear energy contributed less than 1%.

While Pakistan made substantial progress in reducing oil's share in the total energy balance, rising demand offset much of the increase in hydel and gas supply. Moreover, although Pakistan's efforts to increase the domestic production of oil showed success in the commissioning of the Tando Alam and Dhurnal oil wells in the early 1980s, the value of fuel imports continued to rise until 1982,[9] having first jumped from $78.2 million in 1973 to $238.8 million in 1974 and to $681.9 million in 1979. The second spurt in international oil prices in 1980 caused the value of fuel imports to jump to $1,442,100,000 in 1980, then to a peak of $1,616,800,000 in 1982. Thereafter, the value of fuel imports declined steadily, dipping to $1,433,200,000 in 1985 as oil prices continued to fall rapidly in the mid-1980s.

The collapse of OPEC unity on oil pricing policies from 1985 onward and the oversupply of oil available on the international market at substantially reduced prices have not relieved the structural imbalance in Pakistan's terms of trade. The deficit in Pakistan's current account balance, for instance, increased from $753 million in 1981 to $1,110,100,000 in 1982 and reached a depth of $1,268,400,000 in 1985. Part of the reason was the decline in the remittances of Pakistani workers abroad, mainly in the Persian Gulf, which fell from $2,888,200,000 in 1983 to $2,457,100,000 in 1985, although they rose to $2,631,700,000 in 1986. However, as expatriate Pakistani workers continue to return in the late 1980s following retrenchment in the Gulf sheikdoms in the face of falling oil revenues, Pakistan's economic problems are likely to be compounded.

The natural gas sector remains plagued with problems despite the government's supportive production policy, under which the wellhead price per thousand cubic feet (Mcf) of Sui gas rose from 0.75 rupee in 1981, to 5.27 rupees in 1983, and to 30.38 rupees ($0.62) in July 1984. Despite dramatically improved incentives for natural gas producers, prices remain too low to encourage sizable new production. To induce conservation, the government raised consumer prices for gas by 30 percent in January 1982, 23 percent in January 1983, and 25 percent on July 1, 1983. While Islamabad agreed with the World Bank to raise the weighted average natural gas price to two-thirds of the world price equivalent by mid-1988, Pakistani consumer prices remained only about 45 percent of the international equivalent.

In Pakistan, natural gas is used in power generation (21%), in the manufacture of cement (7%) and fertilizer (34%), in general industry (25%), and in the commercial (3%) and household (10%) sectors.[10] Natural gas is obviously vital for power, industrial, and fertilizer sectors of the economy. With demand for natural gas accelerating under subsidized consumer prices, with gas supplies stagnating, and with gas used to make up shortfalls in hydel power arising from drought conditions at Mangla and Tarbela, Pakistan has experienced gas shortages in several parts of the country. During 1983, for example, electricity supply was 12 percent short of demand and the natural gas supply fell short by 14 percent. The shortages, which have at times crippled industrial production in the country, were the major catalyst behind the Sixth Five Year Development Plan, launched in July 1983, giving top priority to energy development, especially natural gas.[11]

While there were many reasons for Pakistan's energy crisis, clearly Islamabad deserves some blame for its management of the energy sector. The state-run Oil and Gas Development Corporation, despite sizable budgetary allotments, has consistently failed to expand oil and gas production to meet the nation's growing energy needs. Similarly, both the Water and Power Development Authority (WAPDA) and the Karachi Electricity Supply Corporation (KESC) have terrible records of shutdowns and breakdowns. During the winter, Karachi—Pakistan's most important commercial, financial, and industrial center—typically experiences power shutdowns of several hours per day. Parts of the Punjab and the North-West Frontier Province have had similar power outages.

While some observers have expressed concern that the recurrent shutdowns of the Karachi Nuclear Power Plant (KANUPP) may be a smokescreen for clandestine diversion of plutonium for a weapons program, the

U.S. government has stated that there was no evidence that Pakistan had reprocessed spent fuel from KANUPP.[12]

In addition to economic losses in the industrial and commercial sectors, power outages have also dealt substantial setbacks to agricultural production for both the domestic and export markets, by disrupting tube well irrigation. Even though the Power Development Authority deserves credit for operational improvements that reversed the upward trend in power and energy losses, annual power losses as a percent of total generation were still 29-30 percent in the mid-1980s, down from about 38 percent in the mid-1970s.[13] While electric generating capacity under the Fifth Development Plan rose from 3,265 Mw to 3,954 Mw by March 1983, it did so only at the cost of 13,000 million rupees. Although Islamabad has made great strides in electrifying the countryside, there are still fewer than 5 million electricity consumers out of a total population of about 100 million.[14]

To cope with the growing shortage of electric generating capacity, variously estimated in the range of 700-900 Mw, the Sixth Five Year Development Plan (1983-88) targeted $7.5 billion out of $36.83 billion for investment in energy.[15] One key priority is to develop 1500 Mw of additional capacity as rapidly as possible to narrow the growing power gap.[16]

In addition to developments in the hydro, gas, oil, and nuclear sectors, Pakistan is developing renewable sources of energy, such as biogas and solar or windpower, to aid those regions where it is prohibitively expensive to extend the national electricity grid.[17]

ENERGY AND THE ECONOMY

While the Pakistan economy has been making great economic progress, the problems are acute—a widening gap between energy supply and demand, low productivity, poor management of government undertakings, and growing alleged irregularities of the earlier Zia regime. In the wake of the devastation created by the Bhutto government's socialization of the economy in the 1970s, the oil price rise, and the loss of East Pakistan, the nation's economic problems have grown in magnitude and complexity.

While Pakistan made great strides during the five year plans of 1978-83 and 1983-88, even the government's assessments have been somber: "It appears, by and large, [that] personal consumption has been sustained at a reasonable level. . . . Poverty was not eliminated, but its frontiers were pushed back from a pervasive phenomenon to identifiable areas to be attacked. But, the situation was much less satisfactory in relation to avenues of social consumption."[18]

While the economic record of the Fifth Plan (July 1, 1978-June 30, 1983) was mixed,[19] clearly the energy sector was the paramount obstacle to sustained ecnomic progress. However, despite setbacks in energy, by the last year of the plan (July 1, 1982, to June 30, 1983) substantial progress had been made, particularly in view of continued world recession. In the 1980s, GNP growth rates have remained above 6 percent.[20] There also was considerable price stability, with inflation decreasing from the average of 10.3% in the period 1965-1980 to an estimated 7.5% in the period 1980-86. Despite some of the structural deficiences in the Pakistani economy, the high rate of economic growth has accelerated demand for energy.

Even though the Sixth Plan gives the highest priority to developing the nation's energy resources to meet the needs of economic growth, Pakistan's problems in the energy sector will not be easy to overcome. The plan called for total installed generating capacity to be increased from 4,809 Mw in fiscal 1983 to 8,604 Mw in fiscal 1988, with most of the new plants to be thermal. However, it appears likely that power shortages will retard Pakistan's economic growth throughout the rest of the decade. With demand for petroleum products growing at a skyrocketing 11.3 percent per year, during the Sixth Plan oil demand was seen rising from 112,000 barrels per day to 192,000 bpd.[21] The major increases occurred in demand for fuel oil (to back out natural gas in thermal generation), diesel oil, and gasoline. While it appeared that Pakistan would surpass its planned target for domestic oil production, nonetheless oil imports during 1983-88 grew as a percentage of total oil consumption.

The dire trends in oil demand are matched in the natural gas sector. Sixth Plan projections showed total gas demand escalating from 702 million cubic feet per day to 839 mmcfd on average, with demand peaking between 979 and 1,121 mmcfd. With domestic supplies forecast to increase only from 622 mmcfd to 825 mmcfd on average and from 692 mmcfd to 916 mmcfd during the peak period,[22] Pakistan was to experience sizable shortages of gas throughout the plan.

The impact of energy shortages on Pakistan's economic, social, and political systems has led to a major policy debate in the country over two alternatives for Pakistan: pouring money into nuclear power to help alleviate growing electricity shortages or moving to develop its vast coal reserves (640 million metric tons). While, as noted, Pakistan's coal reserves are of very poor quality, many energy experts believe that Islamabad should develop the coal sector in stages, with power stations at the production facility, rather than choose the "quick fix" nuclear option.

The country's attempts to acquire nuclear generating capacity have been hampered by its refusal to place all its nuclear facilities under full-scope safeguards. There are growing indications that Pakistan may have embarked on a clandestine nuclear weapons program.

Nuclear Energy and External Security

Perhaps no issue in U.S.-Pakistani relations has been more clouded with
ambiguity and rancor than the development of Pakistan's nuclear energy
sector. From the Pakistani perspective, nuclear energy has, since at least
the mid-1960s, been viewed as a valuable energy source that would have to
be pursued to ensure that the nation met its energy requirements. Both
before and after the oil price shocks of 1973-74, this view was supported by
teams of specialists from the International Atomic Energy Agency.[24]
Critics attacked the IAEA assessments as biased in favor of expanding
power generating capacity rather than adequately assessing how much
demand might be met by reducing power theft and transmission losses.
Similarly, the IAEA's post-embargo analysis of 1975 was viewed skeptically
for assuming a continuation of Pakistan's historically high levels of growth
in power demand, without adequately evaluating the possible impact of
worldwide recession and high oil prices.

Pakistan's severe lack of fossil fuel resources and the shortage of hydroelectric resources near the major electrical load centers could be used to
justify a sizable nuclear power program. However, Islamabad did not
promote a major commercial nuclear power program until after the May
1974 Indian nuclear detonation.

As early as the mid-1960s, U.S. officials became concerned about the
sharp synergisms behind the Indian and Pakistani nuclear programs. At
this time, Pakistani officials voiced their concerns to the United States that
under the rubric of its nuclear energy program India might be moving
toward a nuclear weapons capability. The other side of the coin is that the
Pakistan Institute of Science and Technology had decided as early as 1961
to establish a laboratory-scale "hot-cell" reprocessing plant. While this
facility was not built for over a decade, owing to disagreement in the
Pakistani government about its location, Islamabad's interest in it generated concern in New Delhi.

Indian concern was shared by nonproliferation advocates in the United
States, especially when in 1965 rumors circulated that Ali Bhutto had tried
to get President Ayub Khan to allocate $30 million to build a plutonium
reprocessing plant. While the evidence to support this contention remains
sketchy, Islamabad certainly was aware that in early 1965 India had commissioned a plutonium reprocessing plant. Indian protestations that the
facility was for its civilian nuclear program and not for weapons development were viewed skeptically by Pakistan, especially as strategic analysts
in India were campaigning for acquisition of the bomb in response to
China's 1964 detonation.

Pakistan's refusal to sign the partial test-ban treaty of 1963, despite

India's signing and ratification, inflamed Indian public opinion. Bhutto's 1965 statement that "if India builds the bomb we will eat grass or leaves, or go hungry, but we will get one of our own" did little to assuage extremist Indian opinion. In *The Myth of Independence* (Oxford University Press, 1969), Bhutto made clear that Pakistan had at least given thought to a nuclear deterrent.

The motivations behind India's and Pakistan's nuclear ambitions have been thoroughly delineated.[25] What is important for the purpose of this analysis is Pakistan's rethinking of its defense policy. The complex geographic contours of Pakistan's eastern border with India and the disproportionate size of India's military forces and its territory had led Pakistani strategists to adopt an "offensive defense" doctrine, which prevailed in the 1965 and 1971 Indo-Pakistani wars. Under this Pakistani policy, an early advance could secure strategic gains to be held until intervention by the international community effected a ceasefire. In an ensuing settlement, Pakistan, having headed off a lightning strike by India, might gain some small additional strategic advantage.

While this policy may have made some theoretical sense as a broad strategic doctrine, the disastrous 1965 and 1971 wars, combined with India's de facto acquisition of nuclear arms in 1974, led to a reassessment of Pakistan's policy. After 1974, as Stephen Cohen notes, Pakistan's growing conventional military inferiority, its worsening internal problems, and the limitations on its historical allies led to a fundamental reassessment of the country's strategy.[26]

Pakistani fears of Indian designs on its territory can only be understood if viewed against Indian support for Bangladesh's secession in the 1971 war and continued provocations by New Delhi against the Moslem majority population in Kashmir. The 1971 Indo-Pakistani war and the creation of Bangladesh marked a watershed in relations between the two nations by substantially altering the balance of power between them. Given the devastating dislocations in Pakistan's economy caused by the loss of Bangladesh, the intensified Indian expansionism after 1971, and the staggering impact of the 1973-74 oil price rises and subsequent energy shortages, Pakistani government planners had more than enough justification to believe that their national security was endangered on all fronts. Some elements of the Bhutto government believed that only a small nuclear program would serve to deter an Indian attack.[27]

In the early 1970s, Iran's emergence as the major military power in the Persian Gulf region aggravated its longstanding rivalry with India and drew Pakistan into closer association, as a result of Pakistan's conviction that India wanted to dominate the region from the Persian Gulf to the South China Sea.[28] India's interventions in Sri Lanka in 1971 and 1987-89, its annexation of Sikkim in 1973 and intervention in Maldives in 1988, as well as the signing of the 1971 Indo-Soviet Friendship Treaty and the 1978 Afghan coup, served to accentuate Pakistan's perception of India as an expansionist power and of its own diplomatic isolation.

Pakistan's first nuclear reactor at Karachi went critical in 1971. While both the KANUPP power reactor from Canada (1965) and the PINSTECH research reactor from the United States (1961-62) were acquired on a turnkey basis, it had been clear all along that in time, the Pakistanis would take over the reactors themselves.[29] Despite the lack of a formal announcement concerning expansion of its nuclear power program, circumstantial evidence indicates that by late 1971 Pakistan was considering the acquisition of nuclear weapons. In addition to Pakistan's continued warnings to Washington about the real intent of India's nuclear program, and Pakistan's refusal to sign the Non-Proliferation Treaty in the absence of an Indian signature, there is some indication that Pakistan commenced discussions with a French firm, Saint Gobain Techniques Nouvelles (SGN) for the design of a pilot-scale reprocessing facility to be used at PINSTECH for experimental research.[30]

Although there is limited nonclassified evidence to corroborate either that such discussions in fact took place or that in 1973 a tentative contract for the design work was concluded, pending the completion of financial arrangements, this author was told by high-level sources in Pakistan and the United States that such did occur.[31] However, when a team of nuclear personnel from Oak Ridge National Laboratory visited Pakistan in April 1972, key officials of the Pakistan Atomic Energy Commission (PAEC) indicated no interest in accelerating Pakistan's nuclear power program, although by that time Pakistan had 550 qualified nuclear engineers and scientists.[32]

In their report, the U.S. team observed that Pakistan's nuclear power program was moving ahead slowly, with the next reactor (350 Mw) not scheduled to come onstream until 1979. The report noted that the PAEC had not yet chosen the reactor site or design type and had only tenuous plans for nuclear power development after 1979. It specifically stated that no member of the PAEC organization had expressed any interest in obtaining fuel reprocessing technology or other "full fuel cycle"—meaning uranium enrichment—facilities.[33] Despite such official statements, Dr. I. H. Usmani said in March 1972 that he had been forced out as chairman of the PAEC because he had refused to build a bomb.[34]

In June 1973, Islamabad announced approval of a plan for nuclear desalination of sea water and expansion of its nuclear power program. While details were scarce, in July 1973 the government announced that a

500-Mw reactor would be built in the northern part of the country within two and one-half years.[35] This goal proved to be extremely optimistic; budgetary allocations for the reactor, which in the end was slated for 900 Mw, were not made until after the Indian nuclear detonation, and to date the planned reactor has not been constructed.

Although it is difficult to determine the exact date in 1973 Pakistan began discussions with France for purchase of a commercial-scale reprocessing facility, there is little doubt that the negotiations were launched prior to the October 1973 OPEC oil price increases.[36] The timing contradicts Islamabad's claim that the reprocessing negotiations arose out of concern, generated by the energy crisis, about secure access to nuclear fuel supplies. This claim is also inconsistent with the PAEC's December 1973 announcement that "abundant quantities of uranium" had been found in south Punjab and at three other locations in Pakistan. If proven domestic uranium reserves were adequate to support a major expansion in the number of commercial nuclear power reactors, there would be little incentive to acquire a reprocessing facility. As it turned out, however, Pakistan had not, contrary to official reports, found "abundant" quantities of uranium.

Of equal importance is the question of whether the French-Pakistani negotiations over a commercial reprocessing facility preceded alleged negotiations for a pilot-scale facility or occurred in tandem—or whether both moves were subterfuges designed to draw attention away from Pakistan's clandestine development of a uranium enrichment facility. From recent reports, it is clear that the pilot-scale reprocessing deal was consummated and Pakistan acquired the plant sometime during 1975-76.[37]

Recent attention in the West has been riveted on the implications of Pakistan's acquisition of a uranium enrichment plant capable of making nuclear weapons; from the evidence available, however, it appears that even prior to late 1976 or early 1977, Pakistan intended to keep open the weapons option by acquiring the pilot-scale reprocessing facility. Since Pakistan's one nuclear reactor was of CANDU design, the acquisition of even a pilot-scale reprocessing facility would have allowed Pakistan to extract enough plutonium to make a small number of bombs over a period of three to four years. Moreover, if Pakistan acquired a commercial-scale reprocessing facility, it could secure much valuable technical expertise directly applicable to a clandestine weapons program.

While it is impossible to prove without access to classified information, it appears, on the basis of extensive interviews in France, the United States, and Pakistan, that throughout most of 1975-76 the Bhutto government did not have the intention of directly manufacturing a nuclear

weapon. Pakistan moved to purchase the equipment for a uranium enrichment plant only after it became apparent that the country's scientists
lacked the technological know-how for making nuclear weapons via the
reprocessing route and that the commercial reprocessing facility would be
delayed indefinitely. By that time, profound changes had occurred in
Pakistan's geopolitical situation.

In addition to seeking reprocessing capability, the PAEC in 1973-74
moved to acquire a fuel fabrication facility from Canada to supply the
KANUPP reactor in Karachi. It appears that Pakistan also tried to purchase a heavy water production plant, without initial success.[38] Despite
the various initiatives, on the eve of the Indian nuclear explosion in May
1974, Pakistan's nuclear facilities nonetheless remained primitive in comparison with New Delhi's advances.

PAKISTANI NUCLEAR POLICY AND THE INDIAN NUCLEAR EXPLOSION

While Pakistan's severe lack of fossil fuels and the geographical disadvan-
tage of its hydroelectric resources could easily have justified a sizable
nuclear power program, as late as September 1974 Pakistan officially had
firm plans for only one additional 600-Mw reactor; studies were underway
for a third reactor to be completed sometime in the 1980s. It was not until
December 1974—seven months after the Indian nuclear explosion and
fourteen months after the onset of OPEC's oil price increases—that
Pakistan announced a major expansion in its nuclear reactor program.
Under the new program, Islamabad would build four or five new power
plants in the 1980s and one plant every year during the 1990s. This plan was
subsequently modified to call for construction of twenty-four nuclear
power plants by 2000.[39]

The events that transpired in Pakistan's nuclear policy between 1974
and the overthrow of the Bhutto government in July 1977 have been well-
delineated.[40] By 1978, intense diplomatic pressure by Washington and
Ottawa had not only led to the termination of all nuclear cooperation
between Canada and Pakistan, but had also, under U.S. direction, forced a
cancellation in the French-Pakistani reprocessing agreement.

It is extremely difficult to analyze Pakistani nuclear energy policy
during the 1974-77 period because fears for national security arising from
the Indian explosion have been so pervasive in Pakistani political debate.
The role of nuclear energy, however—in both its peaceful and destructive
applications—is viewed much differently in Islamabad than in Washington. In Pakistan, development of commercial nuclear technology is
perceived as a symbol of a nation's "modernity." While it would be dan-

gerous to emphasize this perception, it would be equally dangerous to dismiss this psychological component of the nuclear debate.

While Pakistan's slow implementation of a major nuclear program in the wake of the oil embargo raised disturbing questions as to the program's motivations, the IAEA in fact supported a major expansion in the Pakistani nuclear program on the eve of the embargo and reiterated this support in 1975. Furthermore, in spite of waning enthusiasm in the industrialized world for commercial nuclear power as an alternative to oil, the Pakistanis found sufficient justification for nuclear expansion in their unique energy situation in 1976: the sizable fluctuations in electric generating capacity, the concentration of hydroelectric capacity in the north while demand is greatest in the south, and the staggering costs of long-distance transmission lines.

Because the nuclear question was part of national security considerations, the Bhutto regime created a major commercial nuclear power program to justify Pakistan's long-term needs for a reprocessing facility. Although the IAEA's assessment of Pakistan's need for nuclear reactors was more modest than the PAEC's projections, the IAEA's projections in both 1973 and 1975 did, within the dictates of the plutonium recycling debate at the time, justify Pakistan's future need for a reprocessing facility. It was for this reason that the IAEA board of governors approved the French-Pakistani reprocessing transfer and concluded the trilateral safeguards agreement.

While reasonable analysts may differ over whether Pakistan needs more electric power generation plants (nuclear or otherwise), or whether the country would be better advised to use scarce resources to improve the transmission and distribution system, there are legitimate reasons for advocating nuclear power expansion. However, Pakistan's preoccupation with national security and the defensive role of nuclear weapons raises vexing problems. In responding to Washington's opposition to the reprocessing facility, Islamabad relied largely on the argument that it is not the mere possession of sensitive nuclear technology that threatens international security, but rather a nation's perception of being so overwhelmingly threatened that it has no recourse but to acquire nuclear weapons. While sophisticated Western-educated analysts might not understand this view, failure awaits any U.S. government policy that fails to take into account Pakistan's perception of Indian intentions.

Pakistan has traditionally sought security through membership in regional alliances, such as CENTO and SEATO, that are backed by the United States, and through close relations with Iran and China. While the specifics of its foreign policy have changed over time, Pakistan has tried to ensure sound relations with at least one of the major powers. In the two major crises of Pakistan's short history—the 1965 and 1971 wars—the United States failed to prevent catastrophic defeat. Surprisingly, very few Pakistanis are bitter over this fact. However, as a long-standing ally of the United States, Pakistan believes that Washington should reexamine the "entire context of Pakistan's security options and recognize the threat Pakistan feels from its neighbors."[41] In Islamabad's view, the United States should either make diplomatic initiatives to protect Pakistan from India and Afghanistan (meaning the Soviet Union) or, at the very minimum, not stand in the way of a firm Pakistani response to these threats.[42]

In Islamabad's view, the Carter administration not only turned a deaf ear on Pakistani security concerns but also made a conscious decision to "tilt" U.S. policy toward India. A series of events—President Carter's rejection of Pakistan's request for 110 Vought A-7 attack planes, the resumption of shipments of enriched uranium to India, the inclusion of India (and the exclusion of Pakistan) on President Carter's December 1977–January 1978 world tour, the failure of the United States to step up military assistance to Pakistan in the wake of the May 1978 Afghanistan coup, and Pakistan's deteriorating security situation in the wake of the Iranian crisis—all combined to convince Islamabad that it could not count on the U.S. commitment to Pakistan's security.

THE ZIA GOVERNMENT: ESCALATION IN NUCLEAR TENSIONS

While Bhutto charged that the United States had assisted in his overthrow for a quid pro quo that the Zia government would cancel its reprocessing contract with France, in reality the situation was far more complex. While initially there was hope that the new Zia regime might cancel the SNG-PAEC reprocessing agreement, it did not. On numerous occasions, the chairman of the Pakistan Atomic Energy Commission and the federal minister of science and technology reaffirmed that the deal would be consummated.[43]

The reprocessing controversy continued after 1977, with rumors abounding that China or Libya was financing or otherwise aiding a nuclear weapons program for Pakistan. Following the Communist-backed April 1978 coup in Afghanistan, however, new strategic concerns in the Middle East and South Asia began to cloud the direction of U.S. nonproliferation policy. Although the strategic motives and the role of the Soviet Union in the series of Afghan coups that occurred between April 1978 and December 1979 remain subject to intense debate,[44] the April 1978 coup in Kabul added a new dimension to the geopolitics of the region and made U.S.-Pakistani relations far more complex.

In the immediate aftermath of the coup, Islamabad attempted to warn the United States that not only was the Soviet Union behind the events in Afghanistan but also that it represented the first component of a Soviet thrust against the Persian Gulf. The United States chose to dismiss Islamabad's concerns as merely a subterfuge designed to reduce U.S. diplomatic pressure against Pakistan's clandestine nuclear weapons program. Ironically, the United States apparently discounted Pakistan's warnings largely because Washington believed the Shah of Iran could deal with any Afghan-supported efforts to subvert the Gulf states.

To be fair to U.S. government leaders, in April 1978 there appeared to be little prospect that Moscow was preparing to intervene in the Gulf. From Washington's perspective, alleged Soviet involvement in the overthrow of the Daoud regime was largely designed to restore Soviet influence in nonaligned Afghanistan and to reverse the country's growing drift into the Iranian political orbit. While Washington was clearly concerned about the direction of events, the overthrow of Daoud was not perceived as a fundamental shift in the geopolitics of the region. This perception of events was not shared by Islamabad or by many other observers of the Afghan scene. [45]

Although unknown at the time, the French decision to cancel the reprocessing contract with Pakistan did not disturb Islamabad, for two reasons. First, despite the cancellation, France, apparently for motives still unknown, transferred 95 percent of the blueprint for the facility. Second, Pakistan in 1977 apparently established a number of dummy corporations around the world to purchase the component parts of a centrifuge uranium enrichment plant. [46] Abdul Qadir Khan, a Pakistani scientist, is believed to have stolen the designs for the enrichment plant while working at the Almelo plant in the Netherlands.

These events, which alarmed Washington far more than the political situation in Afghanistan, came to light in September 1978 when Frank Allaun, a British Labour M.P., received information regarding Islamabad's purchase of centrifuge equipment in the United Kingdom. Allaun notified Energy Minister Anthony W. Benn, who in turn alerted the German and U.S. governments. In March 1979, the CIA informed President Carter that Pakistan was indeed building a centrifuge plant capable of producing weapons-grade uranium. In April, the United States, invoking the Symington amendment, announced that, as a result of Pakistan's construction of a uranium enrichment facility, it was canceling all military assistance and would reduce its economic aid. In June, Washington announced that Pakistan would receive only $40 million of its $120 million request.

In retrospect, it is curious that at the same time the U.S. was curtailing military assistance to Pakistan out of concern that Islamabad's nuclear weapons program would destabilize the subcontinent, Pakistan was moving to improve relations with its arch adversary—India. In May 1979, Pakistan's foreign secretary visited New Delhi to discuss bilateral consultations on problems of regional and international concern. Puzzled by the growing U.S. restrictions on foreign assistance, Islamabad withdrew from CENTO and embarked on a crash program of Islamization. Chief among Pakistan's reasons for withdrawing into nonalignment was a desire to attract greater economic assistance from oil-rich Islamic nations.

At the same time that Pakistan was moving to improve its relations with India and strengthen ties to the Islamic world, it attempted to keep good relations with the Kabul regime after the April 1978 coup. Pakistan was the first state to recognize the Afghan regime of President Taraki, and President Zia made an early state visit to Kabul to exchange views with the new regime. As a further demonstration of good faith, Islamabad moved to facilitate the transit of Afghan goods through Pakistan. While all this was going on, Pakistan maintained its historically close relations with Peking and Teheran. Yet, despite Pakistan's astute regional diplomacy from mid-1978 to mid-1979, events in the region conspired against Islamabad's vital interests. In January 1979, the Shah of Iran, the chief military bulwark against the fissiparous tendencies of Baluchistan and the North-West Frontier Province, was overthrown and replaced by the militarily weakened Khomeini regime. Recurrent coups and coup attempts in Afghanistan led to civil war and the migration of thousands of Afghan refugees into Pakistan.

The influx of refugees posed serious economic problems for Pakistan, and Baluchistan and the North-West Frontier Province became bases for arming and training Afghan insurgents. Islamabad's tenuous military control of these two regions gave rise to worries that aid to the refugees might bring retaliation from the Soviet-backed Kabul regime. Alternatively, Islamabad feared that the weapons pouring into the region might be turned against it once the regime in Afghanistan was toppled.

In November 1979, relations with the United States deteriorated badly when a coalition of right-wing Muslim students, leftist Iranians, and Palestinians burned the American embassy in Islamabad, incited by allegations of American participation in the seizure of the Grand Mosque in Mecca.

In December 1979, Pakistani security was further jeopardized when the Soviet Union, concerned about the deterioration in its political position in Kabul, invaded Afghanistan with massive land and air forces. The Soviet intervention sparked a new flight of refugees to Pakistan, and Pakistan's relations with Kabul and Moscow spiraled downward. Given the Soviet claim, supported by India, that its intervention was a consequence of outside interference in Afghanistan's internal affairs, Islamabad was concerned that Moscow and New Delhi might use the pretext of Pakistan's support for the Afghan rebels as justification to dismember Pakistan. In reaction, Pakistan intensified its efforts to acquire nuclear weapons to defend its territorial integrity.

The fall of the Shah in January 1979, combined with the Soviet intervention in Afghanistan in December, led to a dramatic short-term improvement in U.S.-Pakistani relations. Relations again cooled, however, when President Zia rejected President Carter's offer of a $400 million aid package as "peanuts." While Pakistan wanted to bolster relations with Washington out of concern that Moscow might use force against it, President Zia could not risk a closer association with the United States unless Pakistan received enough assistance to dramatically improve its defense capabilities vis-á-vis India and Afghanistan.[47]

Given the political perceptions of the Reagan administration about the Soviet Union's grand strategic design, Washington agreed in 1981 to give Islamabad $3.2 billion over five years in military and economic assistance, making Pakistan the largest recipient of direct U.S. aid outside Israel and Egypt. While Islamabad pledged that it would not pursue a nuclear weapons program, it has nonetheless apparently violated that promise.[48] In any case, the Reagan administration continued to view close U.S.-Pakistani ties as a vital link in countering Soviet military moves in Southwest Asia. Pakistan is not only well positioned as a base for protecting the Persian Gulf but also has highly trained armed forces, among the best in the region. As an entrepot for a rapid deployment force to protect allied access to the Persian Gulf oil fields, Pakistani bases would be a vital strategic asset. In light of these considerations, the U.S. Congress on April 27, 1981, granted an unprecedented country-specific nonproliferation waiver for aid to Pakistan, despite repeated allegations by both European and American sources that Pakistan had embarked on a nuclear weapons program.[49]

Despite numerous disclaimers by both the Pakistani and U.S. governments that Islamabad has no intention of manufacturing nuclear warheads or acquiring nuclear weapons, critics of U.S. policy toward Pakistan are alarmed by a series of actions that they believe contradict such assertions:[50]

- In December 1980, Canadian officials in Montreal seized U.S.-made electronic equipment being shipped illegally to Pakistan. The equipment included components for an inverter that could be used to enrich uranium to weapons grade. Canadian officials at the time said there was evidence of at least five other shipments of similar electronic parts.

- In October 1981, customs agents at John F. Kennedy International Airport in New York seized 5,000 pounds of zirconium illegally bound for Pakistan. Export of this metal is regulated because of its potential use in nuclear operations.
- In 1981, Pakistan obtained a fuel fabrication plant with the help of West German companies.
- For at least eight years, there have been allegations that Pakistan has received Libyan help in obtaining "yellowcake," or uranium concentrate, from Niger, that it has received financial assistance from Libya and/or Saudi Arabia, and that it has been given technological guidance by China. In 1984, rumors of Chinese-Pakistani cooperation delayed the signing of a U.S.-Chinese agreement for nuclear cooperation.
- In July 1984, two Pakistanis were convicted in Canada of illegally exporting U.S.-made equipment that could be used in a nuclear plant. The company the men worked for, Serabit Electronics, was also convicted.
- On July 21, 1984, the *Washington Post* reported the indictment in Houston, Texas, of three Pakistanis for attempting to export illegally high-speed switches called krytons that can be used in building a trigger for a nuclear bomb.
- On July 11, 1987, U.S. customs officials in Philadelphia arrested a Pakistani-born Canadian who was attempting to export to Pakistan 25 tons of a special steel used in uranium enrichment centrifuges and an undisclosed quantity of beryllium, for which one application is increasing the yield of nuclear weapons.

Without access to classified information, it is impossible to assess whether Pakistan has a dedicated nuclear weapons program, but, there can be little question that Pakistan could if it so desired commence a weapons program on a limited scale, perhaps five to seven bombs per year.[51] These weapons would be of a more advanced level than the early bombs produced in the United States.

Unanswered questions are cause for concern about the direction of Pakistan's civilian nuclear program. First, why does Pakistan need commercial reprocessing when it is clearly uneconomical in the absence of a massive civilian program, which Islamabad does not even have on the drawing boards? Second, how can the outside world cavalierly dismiss concern about Pakistan's nuclear intentions in the light of statements by former Prime Minister Bhutto, A.Q. Khan, and others about the need for a nuclear weapons program? Third, why did the IAEA, an impartial agency that has consistently supported Pakistan's need for nuclear power, express concern in 1981 that diversions of plutonium-bearing fuel from the KANUPP reactor may have occurred? Fourth, what did U.S. intelligence satellites detect in the Baluchistan mountains in early 1981? If it was not a nuclear test site, as alleged, what was it? Fifth, why are IAEA monitoring cameras and other safeguard devices subject to chronic operational failures? Sixth, why does Pakistan need an uranium enrichment facility when its only reactor works on natural uranium? Why does Pakistan need a reprocessing plant when it has no breeder reactor program?

ENERGY AND SECURITY: THE PROSPECTS

The fierce debate about nuclear weapons has beclouded the question of whether Pakistan in fact needs a civilian nuclear power program. The answer depends upon a host of policy considerations, the most important being the structure of the gross domestic product–to–energy coefficient, especially in the electricity sector; the relative costs/benefits of constructing electric generating facilities versus upgrading the power transmission and distribution infrastructure and/or reducing power theft; the impact of subsidized consumer and producer prices in skewing the pattern of energy demand and supply; and the difficulty of obtaining financing for energy projects that are not "show pieces" even though more effective on a cost/benefit basis.

Analysts who argue that under no circumstances does Pakistan need civilian nuclear power must find alternative methods to meet the unique problems plaguing the electric power sector. Critics must also take into consideration both the IAEA's assessment of Pakistan's need for nuclear power and the interest of countries such as France, West Germany, Belgium, and Switzerland in making nuclear sales to Pakistan if adequate safeguards arrangements can be effected.

Until the problems and opportunities of Pakistan's entire energy sector are evaluated authoritatively and a coherent resource policy is implemented, a detailed analysis of the role of nuclear energy in the nation's future is impossible.[52] While the author believes that more attention should be focused on alternatives to new large-scale electric power generation—including conservation, improvements in power transmission and distribution, more effective management and accountability practices in the national economic and energy bureaucracy, better energy planning mechanisms between Islamabad and the provinces, and the end of price controls on all fuels—he is critical of those analysts who make facile judgements that under no circumstances is there a role for nuclear power in Pakistan's energy future. Likewise, he queries whether large-scale

high-cost alternative programs such as the Lakhra coal project have not
been offered as much for political, nonproliferation purposes than because
they make sound economic or technical sense.[53]

The critical link between Pakistan's energy and security policies is not
the possible use of its civilian nuclear energy program as a cover for
developing weapons. Rather, by weakening Pakistan's already fragile economy, the energy crisis exacerbated Pakistan's internal and external security problems. The fall in oil prices to about $13 per barrel in 1988 has not
eased security problems but has added new economic problems with the
return from the Persian Gulf of large numbers of Pakistani workers whose
petro-dollar remittances to Pakistan will be lost.

While the magnitude of political and economic crises confronting
Pakistan makes concern over energy issues seem somewhat parochial, in
reality many of the energy issues are but another manifestation of the
country's political problems: the regional imbalance in economic development, the disparity in the quality of life between rural and urban areas, the
tensions engendered in the transition from a traditional to a modern
economy, and the overemphasis on the industrial at the expense of the
agricultural sector. Rising dissidence effected at least in part by economic
dislocations created a perception among some segments of the Zia government that the country was under siege.

While relations between India and Pakistan have waxed and waned over
the last 10 years, Islamabad's diplomacy has enjoyed some notable successes. Relations with Iran have improved, as evidenced by rising volumes
of trade. Pakistani-Chinese relations remain excellent. The 170 J-6 fighters
supplied by China earlier are now being replaced by the Chinese J-7
fighters, with sixty having been purchased in 1986 and an additional 150 on
order in 1989. In addition, 100 upgraded versions of the Chinese Q-5
ground attack fighters have been sold to Pakistan. Pakistan's relations with
the Arab world have brought increased diplomatic support against India by
the Islamic Conference, substantial foreign assistance, and sizable remittances—although declining in 1989—from Pakistani workers in the Gulf.
In 1989, the withdrawal of Soviet forces from Afghanistan secured Pakistan's western front and brought the prospect that a friendly Islamic government would eventually replace the Marxist government of Najibullah in
Kabul.

Despite major diplomatic problems with the United States over both its
nuclear program and its rising illicit drug trade, the Zia government in
1988 reached a $4.02 billion military-economic six-year agreement with
the United States that included the prospect of sixty F-16 fighters. Pakistan
had received forty-nine F-16s under a similar agreement with the United
States in 1981. Relations between the two countries are likely to improve

even further following the death of Zia in a plane crash in 1988 and the election of Benazir Bhutto through the democratic process.

The nature and magnitude of the U.S.-Pakistani assistance agreements have been criticized by many analysts, who fear dire consequences, particularly as the United States has agreed to equip the transferred F-16s with the ALR69, a top-secret electronic device for detecting radar signals. While the F-16s are not fitted with the sophisticated avionics for launching nuclear missiles, critics of the deal believe that Pakistan can retrofit the planes for this purpose.

The acquisition of the sophisticated F-16s, together with reports of the deployment of French Crotále antiaircraft missiles to guard the alleged enrichment plant at Kahuta, has generated concern that India and/or Israel might launch a preemptive strike against Pakistan's nuclear facilities. To a certain extent, this fear was alleviated in December 1988 when Prime Minister Rajiv Gandhi and the newly elected Benazir Bhutto signed an agreement not to attack each other's nuclear reactors. It is also feared that the combination of the F-16s and Pakistan's nuclear program will be seen by India as sufficient reason to develop a dedicated nuclear weapons program, including thermonuclear devices.

From the vantage point of Washington, Pakistan is a valuable ally not only in countering Soviet aggression in Afghanistan but also in "guarding the back door to the Gulf." Moreover, the United States has intense interest in regaining access to the sensitive intelligence facilities in Peshawar, and in receiving basing rights to new ports on the western edge of the Baluchistan coast. While the question of whether the acquisition of sophisticated conventional forces will enhance or reduce Pakistan's interest in nuclear weapons has been debated since the Nixon Administration,[54] clearly the Reagan administration believed that broad U.S. strategic interests outweighed immediate nonproliferation concerns.

What has been lacking in the debate over U.S.-Pakistani relations in general and U.S. nonproliferation policy in particular is any serious discussion of how the acquisition of nuclear weapons would bolster Pakistan's security. In contrast to most South Asian security analysts in the United States, Pakistani defense planners believe that India now possesses a stockpile of strategic nuclear weapons. They hold that if Pakistan is to guarantee its territorial integrity against India's superior military forces, it too must have nuclear weapons. India's attempts on numerous occasions to abet regional separatist movements in Pakistan and its active supported for Bangladesh's independence are 1971 is cited to support this view.[55]

However, despite the apparent strategic rationale behind this doctrine, it has little or no applicability outside a life-and-death struggle with India. This view is supported by Zalmay Khalilzad, who notes:

Pakistan will force several dilemmas. Its nuclear deterrent will not necessarily be credible against limited Indian conventional military attacks, or military moves in Kashmir. . . . Pakistan would be unable to target significant Indian urban centers without putting very large Muslim minorities . . . in direct jeopardy. . . . Pakistan would have few incentives to develop battlefield nuclear weapons to contest a major conventional invasion. The population density and narrow frontier in the main invasion corridors of Punjab would put its own population at risk from tactical nuclear warfare.[56]

Nonetheless, despite these real limitations on Pakistan's use of nuclear force, there is little reason to be sanguine about the course and direction of Indo-Pakistani relations. As long as India believes that there is a real prospect of nuclear blackmail by China in support of Pakistan, and as long as Pakistan believes that India already has strategic nuclear weapons, there will be no reduction in tensions, with or without conventional weapons. The acquisition of sophisticated conventional weapons by either nation will only accelerate the arms race. If a major destabilization in these conventional forces occurs and is exacerbated by a dramatic event such as a nuclear detonation by Pakistan, a major crisis will ensue.

On balance, the critical question confronting the global community is not whether Pakistan is clandestinely developing nuclear weapons under the rubric of its civilian nuclear power program, but how the United States and the Soviet Union can ensure that their respective allied states do not rush to Armageddon, out of a legacy of suspicion, fear, miscalculation, and historical precedent.

NOTES

1. For a detailed discussion of the energy sector, see Charles K. Ebinger, *Pakistan: Energy Planning in a Strategic Vortex* (Bloomington: Indiana University Press, 1981).

2. *Ibid*, p. 25.

3. International Monetary Fund, *Pakistan: Recent Economic Developments*, Washington, D.C., Dec. 5, 1983; U.S. Department of State Airgram, *Pakistan's Economic Policy*, Washington, D.C., Jan. 25, 1984; Alain Cass, "Why the Government Must Move Quickly on Energy," *Financial Times*, April 19, 1983.

4. Ebinger, p. 17.

5. These figures were drawn from *Pakistan Economic Survey, 1983-84* (Islamabad), and *Pakistan Times*, June 17, 1984. The 1986 figure is from *World Development Report 1988* (New York: Oxford University Press, 1988), p. 242.

6. International Monetary Fund, *Pakistan: Recent Economic Developments*, Washington, D.C., Dec. 5, 1983.

7. *Ibid*.

8. *Pakistan Economic Survey 1982-83*, p. 145.

9. These figures are from *World Tables 1987*, (Washington, D.C.: World Bank Publications, 1987), pp. 342-43.

10. *Pakistan Economic Survey, 1982-83*, p. 148.

11. "Pakistan," *Asia 1983 Yearbook*, Far Eastern Economic Review, Hong Kong, 1984, p. 220.

12. See speech by Ambassador Richard T. Kennedy, Georgetown University, Center for Strategic and International Studies, Washington, D.C., June 28, 1984.

13. *Pakistan Economic Survey, 1982-1983*, p. 153.

14. *Ibid.*, p. 151.

15. "Pakistan," *Asia 1984 Yearbook*, Far Eastern Economic Review, Hong Kong, p. 241.

16. *Pakistan Times* (Lahore), June 3, 1984.

17. See *Pakistan Economic Survey, 1982-83*, pp. 156-157.

18. "Pakistan," *Asia 1984 Yearbook*, p. 240.

19. *Pakistan Economic Survey 1982-83*, pp. 1-98.

20. From *The World Bank Atlas 1987*, Washington, D.C., 1988, and the *World Development Report 1988* (New York: Oxford University Press, 1988), p. 222.

21. U.S. Department of State Airgram, *Pakistan's Economic Policy*, Jan. 25, 1984.

22. *Ibid.*

23. *Pakistan Economic Survey 1983-84*.

24. International Atomic Energy Agency, *Market Survey for Nuclear Power in Developing Countries* (Vienna, 1973); IAEA, *Nuclear Power Planning Study for Pakistan* (Vienna, 1975).

25. Rodney W. Jones, "Nuclear Proliferation: Islam, the Bomb and South Asia," *The Washington Papers* no. 82, Center for Strategic and International Studies, Georgetown University (Washington, D.C., 1981); G. S. Bhargava, *South Asian Security After Afghanistan* (Lexington, Mass.: Lexington Books, 1983); see chapter by R. R. Subramanian and C. Raja Mohan on India and chapter by Shirin Tahir-Kheli on Pakistan in James Everett Katz and Onkar S. Marwah, eds., *Nuclear Power in Developing Countries* (Lexington, Mass.: Lexington Books, 1982).

26. Marvin G. Weinbaum and Stephen P. Cohen, "Pakistan in 1982: Holding On," *Asian Survey* 23, no. 2 (February 1983), pp. 123-32.

27. Ebinger, pp. 85-88.

28. For an espousal of this view, see *Pakistan Times*, Sept. 6, 1975.

29. Shirin Tahir-Kheli, *Nuclear Power in Developing Countries*, p. 265.

30. Ebinger, p. 81.

31. *Ibid.*

32. Oak Ridge National Laboratory (ORNL) Document no. 72-5-50, "Report of Foreign Travel to Pakistan, April 24-May 1, 1972," May 25, 1972, declassified from "Official Use Only" under Freedom of Information Act.

33. *Ibid.*

34. Interview with Dr. I.H. Usmani, New York, Oct. 5, 1977.

35. Zalmay Khalilzad, "Pakistan: The Making of a Nuclear Power," *Asian Survey* 16, no. 6 (June 1976), pp. 580-92.

36. Ebinger, p. 82.

37. International Institute for Strategic Studies, *Strategic Survey, 1981-82* (London: IISS, 1982); Steve Weissman and Herbert Krosney, *The Islamic Bomb: The Nuclear Threat to Israel and the Middle East* (New York: Times Books, 1981).

38. *Asian Recorder*, Feb. 5-16, 1974.

39. *New York Times*, April 23, 1976, "Three Years of People's Government," citing Pakistan Government Press Information Department Release no. 13, Dec. 20, 1974, p. 2.

40. Charles K. Ebinger, "U.S. Nuclear Non-Proliferation Policy: The Pakistan Controversy," *The Fletcher Forum* 3, no. 2 (1980).

41. Ebinger, *Pakistan: Energy Planning in a Strategic Vortex*, p. 105.

42. Zalmay Khalilzad, "India's Bomb and the Stability of South Asia," *Asian Affairs* no. 2 (November/December 1977), p. 102.

43. Interviews, PAEC and Ministry of Science and Technology, Islamabad, September 1976; speech by Mohammed Ahmed Khan, Pakistani Ambassador to the United States, Georgetown Center for Strategic and International Studies, Washington, D.C., Sept. 6, 1978.

44. Selig S. Harrison, "Dateline Afghanistan: Exit Through Finland?" *Foreign Policy*, no. 41 (Winter 1980-81): pp. 163-187.

45. *Ibid.*

46. Zalmay Khalizad, "Pakistan and the Bomb," *Bulletin of the Atomic Scientists*, January 1980, p. 13.

47. Francis Fukuyama, *The Security of Pakistan: A Trip Report*, Rand Corporation, Santa Monica, Calif. September 1980, p. 26ff; see also Stephen Cohen, "Security Decision-Making in Pakistan," Report for the Office of External Research, Department of State, Washington, D.C., September 1980.

48. On numerous occasions, Pakistani leaders have eschewed any intention to develop nuclear weapons. See President Zia's Dec. 9, 1982, speech before the Foreign Policy Association in New York.

49. See speech by Senator Alan Cranston (D.-California) before the United States Senate, Washington, D.C., June 21, 1984.

50. "Nuclear Parts Sought by Pakistanis," *Washington Post*, July 21, 1984.

51. Interviews, Arms Control and Disarmament Agency, Washington, D.C., July 1984. See also assessment by Leonard Spector, *The Undeclared Bomb* (Cambridge, Mass.: Ballinger Publishing, 1988) pp. 120-53.

52. See Ebinger, *Pakistan: Energy Planning in a Strategic Vortex*, pp. 109-36.

53. Interviews with Stone and Webster engineering officials, Washington, D.C., July 1984; U.S. Department of State, June 1984.

54. Ebinger, *Pakistan: Energy Policy Planning in a Strategic Vortex*, pp. 93-95.

55. Rodney W. Jones, *Nuclear Proliferation: Islam, The Bomb and South Asia*, Washington Paper No. 2, Center for Strategic and International Studies, Georgetown University (Washington, D.C., 1981); Zalmay Khalilzad, "Pakistan and the Bomb," *Survival* 21 (November/December 1979): pp. 244-50; Major General K. K. Palit and P. K. S. Namboodiri, *Pakistan's Islamic Bomb*, (New Delhi: Vikas, 1979).

56. Zalmay Khalilzad, *Pakistan: The Nuclear Option*, monograph 10, Final Report no. (49-1)-3747, prepared for U.S. Energy Research and Development Administration (Los Angeles: Pan Heuristics, 1977), pp. 29-38; cited in Rodney W. Jones, *Proliferation of Small Nuclear Forces*, report prepared for the Defense Nuclear Agency (Washington, D.C., 1983).

5

Taiwan

DENIS FRED SIMON

Historically, economic growth has been closely associated with expanded energy consumption. One of the primary consequences of the global energy crisis of the 1970s was the recognition that the management and development of energy policy must be closely integrated with and supportive of overall economic objectives. While the OPEC price increases and the uncertainty of supply affected all nations, developing countries paid a particularly heavy price. Because of the expanded emphasis on manufacturing and the development of capital-intensive industries such as petrochemicals during this period, the Third World proved to be quite vulnerable to the instabilities generated by the crisis.

The global energy crisis caused an appreciable slowdown in overall economic growth in Taiwan. OPEC actions also served to crystallize the price of Taiwan's dependence on external suppliers for its fuels. Even though, admittedly, Taiwan fared much better than some other nations in accommodating rising oil prices, it was increasingly difficult for the country's economy to sustain the OPEC shocks, since, like most nations, Taiwan was caught relatively unprepared for the energy crisis of the 1970s. At the same time, however, the OPEC crisis served as the impetus to a fundamental restructuring of the island's economy—a rethinking of the overall approach to the management of its economy as well as its energy future.

This paper analyzes the sources of energy security and insecurity in Taiwan. It examines Taiwan's current strategy for managing its energy affairs, specifying the tradeoffs between the alternative options available. It also highlights the key threats to Taiwan's overall security, with an eye toward the island's primary areas of vulnerability. Given the possible reunification of Taiwan with the China mainland, the paper concludes by projecting how the island's energy future could be affected by progress or lack of progress in the reunification process.

Between 1961 and 1986, Taiwan's real gross national product grew from U.S.$4.5 billion to almost U.S.$7.2 billion, representing an average annual growth rate of almost 9.0%.[1] Per capita income rose to U.S.$3,784 in

Table 5.1.   Power generation and consumption in Taiwan
             (Millions of kilowatt-hours)

	Total Generation	Consumption			
		Industrial	Residential & Commercial	Other	Total
1961	4,084	2,866	662	556	3,528
1971	15,171	10,668	3,148	1,291	13,836
1981	40,150	28,138	9,310	2,620	37,448
1986	59,031	39,840	13,973	2,937	53,813

Source: Energy Committee, Ministry of Economic Affairs, 1987.

1986 from U.S.$142 in 1961, reflecting an average increase in real terms of 6.8% per year. These figures are a manifestation of the major changes that have taken place in the island's economy over the last three decades. The two most visible influences have been the growing role of exports and the increasing importance of manufacturing in the overall mix of economic activity.

The demands of economic growth, as well as the consequences of that growth, have stimulated substantial expansion in the use of energy, especially as Taiwan moved into heavy industries such as steel, petrochemicals, and shipbuilding in the early 1970s. In addition, as table 1 indicates, household power use has grown steadily, reflecting improvements in the standard of living and the greater availability of consumer appliances to the island's more than 19 million inhabitants. Power generation in Taiwan increased from 4,084,000,000 kilowatt-hours (kwh) in 1961 to 59,031,000,000 kwh in 1986, an average growth of 11.3% per annum.[2] By 1990, one source projects probable electricity demand to be 86,897,000,000 kwh, climbing to 174,513,000,000 kwh in the year 2000.[3] The ratio of the growth rate of the GDP to that of electricity during 1954-86 was about 1.35.[4] The faster growth in power output might lead some to suggest that energy demand is being met without a problem. However, given current projections of future energy demand and economic growth, supply considerations will continue to have weight for the authorities on Taiwan for the foreseeable future, especially since electricity growth since the OPEC crisis has fallen below the average growth rate of roughly the last three decades. Moreover, since a large component of Taiwan's energy supply is met through reliance on imports, its future energy position becomes an even more serious concern because of the island's increasingly uncertain political situation.

Prior to 1961, Taiwan's power generation system was based mainly on hydroelectric sources, complemented by a small thermal generating capability. During the period of Japanese colonial occupation of Taiwan (1895-1945), a modest energy capability was created as part of Japan's efforts to modernize the island's overall economic infrastructure.[5] The bulk of Taiwan's early power development, however, was accomplished with assistance from the United States, which during the 1950s and early 1960s helped provide the funds and technical know-how to build up the power system and related infrastructure to meet the local economy's immediate needs.[6] Yet, as demand for electricity began to grow and suitable sites for additional hydropower plants were lacking, the share of thermal plants in the total installed capacity began to increase quite rapidly. Many U.S. firms, at the encouragement of the Agency for International Development, came to Taiwan to help develop thermal capabilities.[7] Growth in thermal generating capacity was supported by the availability of inexpensive imported petroleum.

In general, Taiwan had been highly dependent on external sources for meeting its growing energy needs. This remains true despite efforts underway to cultivate high-technology industries that are non–energy intensive.[8] The China Petroleum Corporation, which oversees Taiwan's oil development and imports, is mainly engaged in refining imported petroleum, though it does conduct limited oil exploration on and around Taiwan. It operates two petroleum refineries with a capacity of slightly more than 600,000 barrels a day. Its fleet of eleven oil tankers has a total capacity of more than 1 million deadweight tons.[9]

Dependence on external energy sources became especially problematic in the early 1970s, when the OPEC crisis shook the island's economy, injuring such key industries as plastics and petrochemicals.[10] Many questions arose about steady access to fuel supplies irrespective of the cost. The basically unstable mode of world oil supplies and prices after the energy crises of 1973 and 1979 hurt Taiwan's attempts to plan for the future. The impact that a cutoff or serious decline in the availability of foreign oil would have on Taiwan's economy is seen in table 2. In 1986, almost 90% of total energy supplies were imports, with petroleum imports accounting for 55%.

Owing to the sharp escalation of oil prices, Taiwan's expenditures for oil imports have grown rapidly since the early 1970s. This has placed a heavy burden on both the private sector, which has depended on low energy costs for maintaining overall competitiveness, and on the government, which periodically has had to subsidize energy prices to offset the economic impact of rising prices.[11] In 1986, over U.S.$2.2 billion was expended for imported petroleum—equivalent to about 9% of total imports—not an insignificant amount. (See table 3.) The addition of coal imports brings

Table 5.2.   Sources of energy supply in Taiwan
              (Percentages of total oil equivalent)

	1961	1981	1986
Domestic	73.3	14.3	10.4
Coal	56.7	5.1	2.8
Crude oil	NA	0.6	.3
Natural gas	0.7	5.0	2.9
Hydropower	15.8	3.6	4.4
Imported	26.7	85.7	89.6
Coal	NA	10.8	18.6
Crude oil	26.7	66.9	54.9
Nuclear	NA	8.0	16.1

Source: Energy Committee, Ministry of Economic Affairs, 1987.

energy imports to almost 11% of total imports. In Taiwan's case, however, it
is not only cost considerations that are of concern; there is also great
apprehension about the security of supplies.[12] These supply concerns
stem from Taiwan's precarious international political status, the steady
decline in the number of countries that recognize Taipei's claim to be the
official government of China, and the island's almost total dependence on
Middle East suppliers.

    Political leaders on Taiwan have taken three major steps over the last
several years to alleviate its high dependence on imported sources of
energy and on a limited number of suppliers. First, beginning with the
announcement of the "energy policy for the Taiwan area" in 1973,[13] there
has been a concerted effort to develop indigenous sources of energy,
mainly nuclear energy.[14] Efforts have been stepped up to develop offshore
sites for petroleum exploitation, though work has been limited by geological uncertainties as well as by the location of some key sites in areas
contested by the People's Republic of China, Taiwan, Vietnam, and the
Philippines.[15] In 1985, the China Petroleum Corporation authorized
McDermott Corporation of the United States to build a drilling platform
and pipeline near Hsinchu to tap offshore oil and gas reserves. Total
production of the well is estimated at 1 million cubic meters of natural
gas.[16] The second initiative has been an effort to acquire more energy from
alternative suppliers, thereby reducing the island's vulnerability in the
case of political disruptions, such as Arab-Israeli wars and the Iran-Iraq
war.[17] Third, programs have been introduced to improve the efficiency of
energy use. As in most other countries prior to the OPEC crisis, con-

Table 5.3.   Energy security indicators
             (Percentages)

	Oil imports/ energy imports	Value of oil imports/GNP	Value of oil imports/total imports
1961	100.0%	1.36%	7.16%
1971	99.9	1.24	3.18
1974	96.3	NA	10.27
1981	78.3	9.40	21.01
1986	61.3	3.08	8.45

Source: Energy Committee, Ministry of Economic Affairs, 1987.

servation measures were not seriously considered in Taiwan, but conservation efforts at the both national and enterprise level have been stepped up.[18]

However, there is only so far that Taiwan can go in diversifying its energy sources from abroad or reducing dependence on imports. Leaders on Taiwan realize that energy conservation also has its limits, achieving savings at the margins, but not really confronting the core of the supply problem. Conversion of power generation to alternative fuels is the major measure by which the island can reduce its dependence on imported oil. In view of the ever-increasing demand for electricity and the continued uncertainty of petroleum supply, the long-range power development program for Taiwan has moved toward emphasis on nuclear power.[19] In fact, Taiwan has established one of the most successful nuclear programs within the Third World. Unlike the case in several other nations, where national security imperatives have given greater impetus to nuclear energy programs, Taiwan's movement toward nuclear has been driven equally by economic motives.

TAIWAN'S POWER INDUSTRY IN INTERNATIONAL PERSPECTIVE

Since the OPEC crises, Taiwan officials have stressed that development of nuclear power is the key to dealing with the energy problem in the future.[20] According to figures released by Taipower, the state-owned energy company that is responsible for overseeing the island's nuclear power program, it is estimated that the cost of power generation by nuclear units is about one-third less than that for oil-fired units. As of early 1985, nuclear-generated electricity costs were the equivalent of 4.5 cents a kilowatt-hour, compared with 5.2 cents for coal and 6.4 cents for oil,[21] and

Table 5.4.   Sources of Taiwan's imports of petroleum
             (Percentages of total)

Country	1980	1981	1982	1985	1986
Saudi Arabia	40%	41%	48%	37%	35%
Kuwait	38	41	32	18	19
Nigeria	3	3	NA	NA	NA
Oman	2	2	3	NA	NA
Indonesia	3	3	2	8	5
Malaysia	1	1	2	NA	NA
Others*	13	9	13	37	41

Source: Energy Committee, Ministry of Economic Affairs, 1987.
*Includes countries such as the U.S., Ecuador, and Venezuela, as well as spot market purchases.

nuclear costs are forecast to decline even further. Under such circumstances, it has become not only politically necessary to explore expansion of the nuclear power program, but economically wise as well.

In some respects, Taiwan's energy situation is not unlike that of several of the other so-called newly industrialized nations. Starting in the early 1950s, when the economy was basically agrarian-oriented, with a low standard of living and a limited number of skilled workers, Taiwan has built an economy where manufacturing in both heavy and, more recently, skill-intensive industries has grown. Economic progress, however, was accomplished despite a poor raw materials base. While many of developing nations have sought to exploit their indigenous natural resources as a means to earn needed foreign exchange for buying foreign technology and equipment to support industrial development, Taiwan and the other Asian newly industrialized countries (NICs) have had no such alternative.[22]

In addition to its lack of natural resources, Taiwan has had to contend with the problem of its uncertain international status, which has become more critical since 1979, when the U.S. decided to normalize relations with the People's Republic of China.[23] The PRC still claims the island of Taiwan as one of its provinces. In fact, if there is only one opinion shared between Taiwan's ruling Kuomintang (KMT) and the Communist leadership on the China mainland, it is that Taiwan is an integral part of China and should not become an independent entity. Except for a few of the smaller, nonstrategic nation-states in Latin America and the Pacific few countries have formal diplomatic relations with Taiwan. The most notable exceptions are Saudi Arabia and South Korea, both of which are being formally and informally courted by the PRC to strengthen relations.[24]

While most of the industrialized and Third World nations maintain strong trade relations with Taipei, the absence of formal diplomatic relations is often a major constraint as well as an inconvenience. The constraints are mainly the difficulties involved in gaining access to international funds and financing through many of the major international organizations, difficulties in negotiating terms of trade with other nations, and the lack of formal representation in key organizations such as the United Nations. Neither the United Nations nor the World Bank includes Taiwan in official statistics—an omission that is ironic in view of the successful model of development represented by the Taiwan.

Unfortunately for Taiwan, these obstacles have intensified as the PRC has become more active in world politics and economic affairs.[25] Beijing, for the most part, has sought to further isolate Taipei through various forms of political pressure on third countries regarding trade or international/regional organizations such as the Asian Development Bank.[26] The attempts at isolation occurred at the same time Beijing was making overtures to Taipei about peaceful reunification and the applicability of the concept of "one country, two systems,"[27] the formula that was used as the basis for the 1984 settlement of the Hong Kong problem between the PRC and the United Kingdom.[28] Prior to the death of President Chiang Ching-kuo in summer 1988, Taipei's strict policy against negotiating with the Communist government in Beijing also limited the growth of trade with the mainland in areas that could be directly beneficial to Taiwan, especially energy and raw materials.[29]

Nonetheless, over the last several years, trade between Taiwan and the PRC has started to grow. Estimates are that trade reached U.S.$550 million in 1984 and U.S.$1 billion in 1985, and surpassed to U.S.$2.5 billion to U.S.$3.0 billion in 1988.[30] Much of this trade is carried out through Hong Kong, though some direct trade takes place between fishing ships from both sides.[31] For the most part, Taiwan receives herbal medicines and agricultural and mining products in return for consumer products and appliances desired by the PRC. The political authorities on Taiwan have periodically cracked down on this growing trade for fear that the local economy might become too dependent on the mainland market, thereby giving Beijing a source of leverage over Taiwan's economic future. Still, Taiwan businessmen feel that the China market is lucrative and cannot be totally ignored, especially as the West and Japan begin to pursue economic dealings with the PRC.

TAIWAN'S NUCLEAR POWER INDUSTRY

When the first oil shock hit in 1973, Taiwan's economy reacted in ways typical of the reactions in other nations. To mitigate the worsening balance-of-payments position of its economy, Taiwan adopted contradictory economic policies, which severely cut into its export earnings at a time when energy bills were soaring.[32] As a consequence, Taiwan's trade deficit ballooned. The trade balance tumbled from a surplus of U.S.$690 million in 1973 to a deficit of U.S.$13 million in the following year. Wholesale prices increased 40.6% in 1974, while the island's rate of growth dragged at 1.1 percent—a mere fraction of the 10.2 percent annual rate of growth dragged at 1.1 percent—a mere fraction of the 10.2 percent annual rate of growth that had been experienced over the previous ten years.

While the major thrust toward nuclear power occurred during this period, the development of nuclear power had been given extensive consideration since the 1950s and 1960s.[33] Two departments were created within Taipower—the Nuclear Power Committee and the Power Development Department—in June 1955 and May 1963, respectively, to plan the introduction of nuclear power in Taiwan. In fact, site selection for Taiwan's first nuclear energy project, which went on stream in 1978, actually began as far back as 1964. Similarly, various writings on the Taiwan economy during the immediate pre-OPEC period give clear evidence that nuclear power was considered an important component of the island's future energy sources—even though at the time the major transition noted in the Ten-Year Long Range Energy Development Plan (1970-80) was the shift from indigenous coal to imported oil.[34] Along the way, Taiwan met the guidelines of the International Atomic Energy Agency safeguards, as well as working with experts from the U.S. Atomic Energy Commission and several U.S. engineering firms to implement nuclear power generation on the island.

Construction of Taiwan's first nuclear reactor began in 1970. Since that time, nuclear power development has proceeded quite rapidly. Taipower now operates three nuclear power stations, each equipped with a similar set of light water reactors (LWRs) The third plant began operating in 1984 and came fully on stream in mid-1985. Total installed capacity reached 5,144 Mw at that time. A fourth plant was approved in early 1985, to become operational in the early 1990s. Initial projections, contained within the island's Ten Year Economic Development Plan (1980-89), were that by the year 2001, there would be fourteen nuclear plants in operation, generating 51 percent of the island's energy supply—as opposed to the current estimate of 33 percent.[35] However, because of recent concerns about environmental pollution and nuclear safety issues, exacerbated by a breakdown and fire at the third nuclear plant in July 1985,[36] this plan for rapid growth has apparently been reconsidered.[37]

The equipment in each of the Taiwan nuclear plants is American, made by either Westinghouse or General Electric, underlining the island's great

Table 5.5.   Taiwan's nuclear power stations

	Reactor supplier (type)	Turbine supplier	MWe rating	Startup	Architect
Chinsan 1	GE (BWR)	Westinghouse	604	Dec. 1978	Ebasco
Chinsan 2	GE (BWR)	Westinghouse	604	July 1979	Ebasco
Kuosheng 1	GE (BWR)	Westinghouse	951	Dec. 1981	Bechtel
Kuosheng 2	GE (BWR)	Westinghouse	951	March 1983	Bechtel
Maanshan 1	Westinghouse (PWR)	GE	907	July 1984	Bechtel
Maanshan 2	Westinghouse (PWR)	GE	907	Mid-1985	Bechtel
Yenliao 1	?	?	1,000	June 1993	?
Yenliao 2	?	?	1,000	June 1994	?

Source: Engery Committee, Ministry of Economic Affairs.

dependence on the United States for technical and political support for its nuclear energy program. The United States has also been the principal source of financial support for projects, providing, for example, about 25% of the loans for the Kuosheng and Maanshan nuclear installations.[38] U.S. engineering firms have been the primary architects and construction consultants. Taiwan has been happy to integrate its local firms into the procurement and construction process but, unlike Japan or South Korea, has no desire to fully domesticize its nuclear power industry.[39] The United States, along with South Africa, is also a major supplier of uranium to operate the island's nuclear plants.[40]

At the present time, Taiwan apparently does not reprocess any of its irradiated fuels. The United States conducts all of Taiwan's spent fuel reprocessing, although in 1982 Taiwan held discussions with the French firm Cogema on this score.[41] Fuel reprocessing on the island itself has long been opposed by the United States as well as other countries in the Asia-Pacific region, for fear that weapons-grade materials might be diverted from such a reprocessing facility. There were reports in the early 1970s that Taiwan had clandestinely built its own reprocessing station at the Institute of Nuclear Energy Research (INER), the island's top research facility for nuclear-related science and technology. The INER had at one time been associated with the Taiwan military, giving cause for concern in many foreign capitals. According to one analyst, the plant was dismantled as a result of significant pressure from the United States.[42] Indications are that Taiwan officials would welcome participation in some sort of regional, multilateral organization for fuel reprocessing in the Asia-Pacific region, but given Taiwan's prevailing political status, involvement in such a formal consortium might prove impossible.[43] To prevent the storage of spent fuel from becoming a problem for nuclear plant operations, a radioactive waste facility was constructed on Orchid Island, off the Southeast coast of Taiwan, to store irradiated materials from the nuclear reactors.[44]

Given the priorities stated in the current Ten Year Economic Development Plan (1980-89), energy concerns clearly remain a dominant issue for economic and political leaders. In early 1981, the Ministry of Economic Affairs announced its intention to invest almost U.S.$30 billion in energy development to reduce dependence on oil imports.[45] Similarly, a U.S.$25 billion budget was approved for energy development for the 1992-2001 period.[46] The goals for the energy sector fall into three broad categories: 1) reduce the energy elasticity, that is, the percentage increase in energy inputs needed to raise GNP by 1 percent; 2) diversify energy sources, in terms of both energy suppliers and types of energy supplied; and 3) promote energy technology research.[47] While nuclear power development may have peaked, a principal aim is to expand the role of coal as an alternative source to imported petroleum for electricity generation. There remain many unanswered questions about the shift toward coal, mainly concerning the inefficiency of domestic coal production and the price instability in the world coal market.[48]

As a result, despite the setbacks of the last few years, nuclear energy seems to meet Taiwan's needs in terms of reliability, economic factors, investment resources, and security considerations. Foreign technology acquisition is viewed as a principal means to strengthen domestic capabilities and attain greater self-reliance. In 1981, Taiwan's top officials laid out four goals for nuclear power development, reflecting, in many ways, the critical role of foreign technology: 1) formation of a joint venture nuclear energy company with government backing to import know-how in the areas of design, analysis, and quality control methods for nuclear plants; 2) establishment of a nuclear fuel manufacturing plant using imported technology; 3) expanded participation in international nuclear power development projects; and 4) diversification of nuclear suppliers in the areas of fuel, equipment, and technology.[49] In each of these areas, Taipower has made a concerted effort to achieve the stated objectives. Prior to the postponement of the fourth nuclear energy plant, the clearest example of its intentions was reflected in its decision to open up bidding on this nuclear project to non-U.S. vendors such as Framatome of France and Kraftwerk Union of West Germany. By the beginning of 1987, Taiwan ranked among the top ten countries with nuclear generating capability, having produced 25.8 billion kw of power during 1986.[50] It also had one of the most efficient nuclear plant construction records among the industrialized nations.[51]

For over three decades, the specter of an invasion by the Chinese Communist regime has loomed large in the minds of Taiwan's leaders.

Even after the normalization of relations between Washington and Beijing, the United States, along with Taiwan, has continued to view various developments in the once-contested Taiwan Straits area with a great deal of caution. Since the late 1970s and the announcement of the four moderniza- tion program in China, Chinese de factor leader Deng Xiaoping and his cohorts have focused their attention principally on economic moderniza- tion and technological advance, deciding to forgo much of the ideologically informed rhetoric that had characterized the PRC government during the heyday of Mao Zedong. Nonetheless, while domestic issues have occupied their attention, the issues of Taiwan and Hong Kong have not receded in importance for the present leadership. Resolution of both issues is viewed by Beijing as the last step needed to consolidate and finalize the Chinese revolution.[52]

The historical signing of the agreement on Hong Kong between the United Kingdom and the PRC in 1984, under the principle of "one country, two systems," gave the Chinese leadership a significant dose of self-confidence about its ability to use "diplomatic" means to resolve what heretofore had seemed like an intractable problem.[53] With this air of self- confidence, however, has come a new sense of urgency regarding resolu- tion of the Taiwan problem. While the present leadership apparently would prefer to settle the Taiwan reunification question on peaceful terms (and probably hoped to do so while Chiang Ching-kuo was still alive), it will not relinquish its self-declared right to use force to bring an end to the current situation. Reluctance to commit to a nonmilitary solution repre- sents a major stumbling block to any formal negotiations between the two sides. Even though contacts between the island and the China mainland have continued to grow as Taiwan residents, journalists, and businessmen travel to the mainland for short-term visits, the reality is that the two sides appear no closer to achieving reunification. Moreover, the events in China's Tiananmen Square in June 1989, combined with the political crackdown in the aftermath of this event, have further damaged the potential for reunification.

Several major initiatives have been taken directly by Beijing since 1978 to bring Taiwan "back to the motherland." The most prominent was the so- called Nine Principles enunciated by the late Ye Jianying in 1980. The Nine Principles provided a recipe for uniting the country, while claiming to allow Taiwan to retain most of its freedom of action.[54] Not unexpectedly, Taipei rejected this attempt at rapprochement as well as several other overtures from Beijing, pointing, at various times, to the following reasons for its continued unwillingness to negotiate: 1) the legacy of previous negotiations with the Chinese Communists, when it is believed by the KMT that the CCP on two occasions (1937 and 1949) took advantage of the so-called United Front to undermine the position of the Nationalist regime; 2) the experience of Tibet in 1951, when, despite an apparent agreement granting independence of action, Tibet lost all of its real autonomy to Beijing's control; 3) the general attempts by the PRC since 1979 to isolate Taiwan internationally despite making peaceful overtures; 4) the lack of credible guarantees from Beijing or any third party should an agreement be reached and then reversed; 5) the uncertain staying power of prevailing policies in the PRC after the death of Deng Xiaoping; and 6) the economic and political situation on Taiwan.[55]

Of all the issues, the last is perhaps the most relevant and the one over which both the Communist and the Nationalist regimes have the least control. Economically, the standard of living in Taiwan is well above that on the China mainland.[56] The key to the success of the KMT on Taiwan has been its ability to maintain an atmosphere conducive to economic prosperity—even in the times of international difficulties.[57] Politically, the reins of power in Taiwan still belong to the mainlander-dominated political elite, yet persons born on the China mainland who came to Taiwan in 1949 currently comprise only 15 percent of the island's total population. Having been generally locked out of national-level politics until recent changes in the political hierarchy, most of the Taiwanese majority have established powerful institutional bases in business and the economy.[58] This separation between the mainlander-dominated government and the Taiwanese-dominated economy lasted over three decades because of the mutually beneficial "interdependence" that had developed between the two groups. More recently, however, the gradual political liberalization begun by Chiang Ching-kuo has taken on a momentum of its own, leading to the emergence of a formal second political party on the island, the Democratic Progressive Party (DPP). This has given impetus to a remarkable change in the discourse within the political arena, bringing such once-taboo subjects as "Taiwan independence" and "reunification with the mainland" into center stage both in the press and within the once-sacred confines of government institutions such as the National Assembly.

Under such circumstances, if the mainlander-led KMT were to respond precipitously to Beijing's overtures, it would face a difficult time. For one thing, most Taiwanese have few links with the China mainland, and therefore have no strong emotional attachment to the idea of reunification. Second, most native Taiwanese would probably view any deal between Beijing and the current government in Taipei as a sell-out of sorts, since there is a feeling among many local Taiwanese that neither the Communists nor the Nationalists are really concerned about their welfare. Third, the status quo at a minimum is acceptable to most Taiwanese, who would prefer not to become involved in the intrigues between the two sides and would be happy if commercial transactions could be carried out without any political interference or political consequences. At one extreme end of this group are a relatively small minority who publicly favor a politically independent Taiwan, free of the control of both the KMT and the Chinese Communist Party.[59] Were the KMT to make a sudden move, the ranks of the proindependence group might quickly grow, and the results could be highly destabilizing.

At the same time, Beijing has been growing increasingly impatient with Taipei's recalcitrant behavior. There is a sense that, as growing numbers of the KMT leadership leave the scene because of old age or death, it will be more difficult to maintain the integrity of the historical ties between the mainland and Taiwan, especially since during 1895-1945 the island was a Japanese colony, and at no time has it been under the control of a regime in like the current one in Beijing. Additionally, as one PRC scholar has noted, the PRC government sees the KMT not necessarily as a direct threat to its authority, but as a political competitor, thereby serving as an alternative locus of authority for overseas Chinese as well as several groups on the mainland.[60] A case in point would be a number of prominent PRC scientists and athletes who have defected to Taiwan in recent years.

More important, however, in the eyes of Beijing is that Taiwan represents a symbol of foreign intervention in China, the most serious violator being the United States, which interceded in China's civil war to protect the Nationalist government that had withdrawn to Taiwan in 1949. Had it not been for U.S. involvement these last thirty-plus years, Beijing claims, the Taiwan problem might have been resolved. A measure of extent of the PRC's adamancy on this score is its constant admonishment of the U.S., stressing on almost every high-level exchange of visits between Washington and Beijing that the status of Taiwan is a major obstacle to the improvement of Sino-U.S. relations. In November 1988, for example, China accused the U.S. of hindering efforts at reunification by continuing to expand relations with the island and by cautioning Taipei to be careful about forging too many ties too quickly with the mainland.[61]

The U.S. decision to de-recognize Taipei as the seat of the Chinese government and to establish formal diplomatic relations with Beijing on January 1, 1979, created a major security dilemma in Taiwan. One consequence of the decision was the U.S. abrogation of its long-term defense treaty with Taiwan.[62] Moreover, implicit to many PRC leaders in the terms of the normalization agreement was a commitment by the United States to a gradual, albeit steady, reduction in its direct sales of military equipment to Taiwan.[63] While such a formal commitment was never publicly accepted by U.S. officials, PRC leaders have intimated on several occasions that such a commitment was indeed conveyed by Washington. The passage of the Taiwan Relations Act in April 1979 obligated the United States to maintain its previous level of economic and military exchanges with the island without an explicit defense commitment. It thus guaranteed Taiwan a continued flow of military equipment from the United States—although the U.S. decision to recognize Beijing clearly damaged American credibility in the eyes of the 19 million residents of Taiwan.[64]

Taiwan's Defense Alternatives

Upon considering various options in the aftermath of the U.S. de-recognition, Taiwan officials felt they had four basic foreign policy alternatives: 1) do nothing and simply try to ride out the storm; 2) approach the Soviet Union; 3) declare independence; and/or 4) pursue development of a nuclear weapon to deter a possible PRC attack. For obvious reasons, options No. 2 and No. 3 were unacceptable to the leadership, who in spite of the difficulties at the time, remained steadfast in their commitment both to oppose Communism and to ensure that Taiwan remained a part of China. The nuclear option remains a viable alternative, especially since Taiwan has the technical know-how and qualified personnel to construct a nuclear weapon.

Several compelling reasons, however, apparently made this option, at least in terms of a formal declaration, relatively unattractive to President Chiang Ching-kuo and the other KMT leaders. First, Taiwan is a signatory to the Non-Proliferation Treaty and even though it no longer belongs to the International Atomic Energy Agency, it has agreed to abide by all of the principles of these two major institutions. Politically, even though in some circles Taiwan is referred to as an "international pariah," it would not risk further international isolation by developing a nuclear weapon overtly. It is also constrained by various bilateral conventions with the United States that went into effect as part of the Taiwan Relations Act. Should a violation occur, the U.S. could cut off fuel supplies and refuse to reprocess Taiwan's fuel. It is clear that most other nations would follow suit. Second, President Chiang Ching-kuo noted on several occasions that his government would not employ nuclear weapons against Chinese fellow-countrymen, in effect going on the record that Taiwan would not pursue a weapons capability.[65] Third, the military balance in the Taiwan Straits has tended, though not in absolute terms, to be in Taipei's favor in terms of the level of technology and the possession of modern aircraft, thereby making the costs of an attack—even if successful—generally unacceptable to the PRC.[66]

The type of military threat that Taiwan would likely encounter would not be a nuclear attack, but rather a sea blockade or full-scale invasion, by PRC forces, of Taiwan and/or those islands surrounding Taiwan that are currently under Nationalist control—for example, Quemoy. In light of comments made by former Communist party Secretary Hu Yaobang in summer 1985, a sea blockade would appear to be the most likely form of intervention, although in the short term the PRC does not appear to possess the military capabilities to ensure the success of this tactic.[67] Similar remarks were made by Deng Xiaoping a year earlier in discussions with Japanese officials concerning the reunification question.[68] Such a blockade would sever the lifeline of Taiwan: its international trade links, including the flow of imported fuels to the island.

While the PRC does have nuclear weapons, it seems clear to most defense experts that they are primarily directed against the U.S.SR as a deterrent.[69] Of course, Beijing might threaten Taiwan with nuclear attack if it did not acquiesce to current overtures or if the post–Chiang Ching-kuo regime were to declare independence. Taiwan's new leader, Lee Teng-hui, seems unlikely to pursue the latter action, though he is also unlikely to go far beyond the present levels of exchange that have developed through commercial intercourse. Moreover, under the prevailing political climate in China, the leadership has tended to prefer less extremist actions. Beijing continues to offer various benefits designed to woo those living on Taiwan to make contact with the mainland, such as preferential treatment for investors from the island.[70] This is not to say that Beijing would not issue an ultimatum to Taipei, rather to emphasize that one aim of U.S. policy toward the PRC since recognition has been to provide Beijing with a vested interest in behaving in a more moderate fashion in foreign affairs—including pursuit of the solution of the Taiwan problem through peaceful means.

Of course, there are no guarantees on either side of the straits. Both political systems are in the process of major change. For example, even with the legacy of Ferdinand Marcos's downfall in the Philippines and the creeping crisis of political legitimacy in South Korea looming large in the minds of the KMT leadership, it was hard for anyone to predict that the pace of political liberalization on Taiwan would have occurred so rapidly and so intensively. There are also forces that in effect are working against immediate resolution of the problem, such as Taiwan's continued ability to attract foreign investment and advanced technology from abroad. For Taiwan, these factors assure a level of security that was absent in the past. The events in Tiananmen Square aside, the emergence of a pragmatic leadership in Beijing does much to allay fears of adventurist behavior on the part of the PRC. Under such circumstances, Taipei has lacked incentives for pursuing a nuclear option as part of its defense strategy, especially since a nuclear response to any of Beijing's actions would almost certainly bring destruction in the form of nuclear retaliation.

Nonetheless, whether or not it in fact produced such a weapon Taipei might see a great political advantage in the perception that it could do so, if only to deter any extremist action on the part of Beijing. The value of such a "near-nuclear" option, however, would be conditional on the PRC's willingness to believe that the Taiwan authorities would use such a device if severely threatened. While it is hard to envision a scenario in which Taipei would be pushed to take such an action, the reality is that the political value of a "near-nuclear" approach might outweigh the short-term gains in deterrence if Taipei really had a weapon. This seems to be the thrust of the message delivered when a former high-level scientist from Taiwan's military-operated Chungshan Institute of Science and Technology defected to the United States in 1988. The scientist claimed to have been working on a nuclear weapons project at the time of his defection.[72]

Given that a nuclear exchange with the PRC is an unlikely form of confrontation, Taiwan has attempted to strengthen its conventional forces and modernize existing weapons capabilities. One of its immediate responses to the U.S. de-recognition decision was the creation of a "national defense fund" to expand defense research and development,[73] ensure greater technological self-reliance in the military area, and purchase needed weapons from abroad.[74] The Chungshan Institute for Science and Technology has been given additional resources to development improvements in radar, avionics, aircraft, and various types of nonnuclear missiles.[75] Response to a possible blockade has been high on the defense agenda. Moreover, there is an explicit attempt underway to more closely link civilian technological advances with those in the military, to strengthen the industrial and R&D base for defense.[76]

Advanced defense capabilities have become more imperative as a result of additional pressures by Beijing on the United States to halt arms sales to Taiwan. In 1982, Sino-U.S. relations built to a crisis when Beijing refused to accept continued American arms sales to Taiwan. The signing of the so-called Shanghai Communique II seemingly ended the crisis. In reality, however, it merely served to reinforce the differences between Washington and Beijing about what commitments had been made and toward what end. Moreover, the "agreement to disagree" regarding Taiwan, which had formed the backbone of the normalization process that began with the Nixon-Kissinger initiatives, seemed no longer viable from Beijing's perspective. While U.S. resolve to honor the TRA has remained strong, the momentum of the evolving Sino-U.S. relationship has left the authorities on Taiwan with grave doubts about U.S. behavior in the coming years. At the same time, U.S. leverage over the PRC may have decreased as China has pursued what is described as a "independent" foreign policy, refusing to lean to the American side in the strategic triangle of Beijing, Moscow, and Washington.[77]

Apprehensions have grown as discussions between Beijing and Washington focus on the transfer of advanced civilian and "dual-use" technologies from the United States and other Western countries to China.[78] At times, negotiations between the United States and the PRC have also included the sale of military equipment. While the United States sees such sales primarily in terms of helping China deal more effectively with the fifty-plus Soviet divisions on its northern border and while the United States is on record as restricting the sale of items or know-how that could assist any PRC effort to attack Taiwan, the fact remains that it is extremely difficult to sort out the multiple applications of specific technologies.[79] Taiwan officials also have grown uneasy about projected sales of nuclear power technologies to the PRC. Taiwan claims that such transfers of equipment and related technology could be diverted to military use given the PRC's current technical capabilities, adding to the nuclear weapons that the PRC already has.

PROSPECTS AND CONCLUSIONS

Taiwan's ability to preserve the status quo, which at this time seems to be the Nationalist regime's primary goal, depends to a great extent on its success in maintaining economic prosperity and allowing a greater degree of pluralism in the domestic political realm. Energy security is an important part of the economic question. Should the island begin to experience economic difficulties because of the uncertainty of energy supplies or high energy prices that affect export competitiveness, the political authorities might face an internal crisis. In some respects, the likelihood of such an energy-induced crisis has actually increased somewhat as Beijing has been successful in making political inroads in countries such as Kuwait and Saudi Arabia. In the case of the former, China has been the recipient of several large development loans. In the case of the latter, while formal diplomatic recognition has not been forthcoming, China's attempts to play a constructive role in the Third World have not gone unnoticed in Saudi Arabia.[80] Even more critical, and perhaps more ironic, the PRC has apparently sold its CSS-2 ballistic missiles to Saudi Arabia, ostensibly to help the Saudis defend themselves against an Iranian attack; apparently, there are also dozens of Chinese technicians on-site in Saudi Arabia, assisting with the deployment of these missiles.[81]

Taiwan continues to look to the United States for assistance in meeting energy needs. Even in the case of the Mingtan Hydroelectric Plant, a

multimillion-dollar project to build a pumped storage hydropower facility
for the 1990s, Taiwan initially neglected potential suppliers from France,
the UK, and the Federal Republic of Germany in favor of working with the
United States.[82] For further nuclear construction, should it take place, the
U.S. is also viewed as the primary supplier. More recently, as Taiwan's
trade balance with the United States has grown substantially in its favor,
Nationalist officials have even raised the possibility of purchasing crude oil
from Alaska.[83] Such purchases, if allowed, would go a long way toward
reducing the U.S. trade deficit. More important, access to Alaskan crude
oil would help assure Taiwan of uninterrupted supplies of imported petroleum. U.S. officials, however, have been unwilling to lift the ban on
export of Alaskan oil, given their own concerns about future energy
shortages.

Generally speaking, the issue of energy security in Taiwan's case cannot
be divorced from general concerns about overall national security and
nuclear weapons development. Should Taiwan begin to feel increasingly
isolated in international affairs, believing that its "friends" such as the
United States and Japan were succumbing to pressures from Beijing to
curtail their extensive levels of commercial interaction with the island,
Taipei might respond with a knee-jerk reaction. It could be argued that it is
not the mainlander-dominated regime that is the primary source of concern, but rather a post-Chiang, Taiwanese-dominated leadership that sees
the nuclear option as a leverage point somewhere between an explicit
refusal to accept Beijing's overtures and an outright attempt to declare
independence. The political utility of nuclear weapons in this context
could outweigh the costs, even if the immediate reaction of the United
States, Japan, and others might be condemnation.

The challenge for the United States, Japan, and the PRC is to ensure
that Taiwan's leaders see no need—political or military—to pursue a
nuclear option. Washington and Tokyo must continue, publicly and privately, to insist on a peaceful approach on the part of Beijing in return for
American participation in China's modernization through such means as
expanded technology transfers. Bringing China into closer contact with
the West through increased economic exchanges gives Beijing a vested
interest in avoiding extreme actions that might cause the West to cut off
access to its technology and training. Continued adherence by the United
States to the principles of the Taiwan Relations Act is also essential, since
Taiwan is more of a proliferation worry if it is overly apprehensive about its
national security than if it feels it is operating from a position of relative
strength.

Whether Washington or Beijing wants to accept it or not, the United
States is intimately involved in what happens in Taiwan. A precipitous

withdrawal of explicit support by the United States, whether by the government or the private sector, could be extremely destabilizing to the status quo in Taiwan, possibly sending the wrong signals as far as the island's security is concerned. It is the United States, as much as Taiwan's own resolve, that has maintained stability across the Taiwan Straits. Ignoring that critical role risks an extreme reaction on the part of Taiwan and possibly by the PRC as well—which could prove detrimental to the interests of Washington, Beijing, and Taipei all three. Fortunately, in contrast to its past position, the PRC seems to recognize this reality and has in fact encouraged the U.S. to play a more active role in its attempts to reunify the country.

Of course, events on the China mainland could also prove threatening to Taiwan, in both the short and long term. Current dissatisfaction with Taipei's policy of nonnegotiation at a time when self-confidence is declining in China could lead Beijing to exert even greater pressure on Taiwan's trading partners and other international organizations to minimize relations with the island. At the very least, it is not inconceivable to envision a situation in which firms doing business or wanting to do business in the PRC will face criticism by Chinese officials for their investments on Taiwan. The decision to go this route could be made as a result of frustration or Beijing's perception that it was losing whatever leverage it might have regarding the future of Taiwan.

Similarly, while it continues to be in the interest of Beijing to pursue a peaceful settlement of the Taiwan problem, a confluence of events could lead to the use of force. A series of pronounced economic downturns or internal political problems, combined with a serious foreign policy flap with the United States, for example, could lead the PRC away from its generally moderate behavior over the last several years. A succession crisis on Taiwan might also elicit a military reaction by Beijing. In spite of recent changes in the PRC military hierarchy, it apparently continues to be the bastion of conservatism regarding Taiwan. It would not be going too far to suggest that segments of the People's Liberation Army, along with components of China's Ministry of Foreign Affairs, have been the main sources behind much of the pressure on the United States regarding Taiwan.

In the final analysis, resolution of the Taiwan problem depends on a combination of political and economic factors. While the current atmosphere appears to be one of relative calm, especially considering the often-sharp rhetoric that has been exchanged by Beijing and Taipei at times since 1949, the future is anything but certain. Beijing seems to be counting on the sudden downturn in the Western economies and the rise in protectionism since the early 1980s to be the very catalyst needed to drive a wedge between Taiwan and its Western supporters. Under circumstances of dwindling access to Western markets, Chinese leaders contend that Taiwan's business community will increasingly be driven to seek out economic opportunities with the China mainland. Such an expansion of trade—with energy sources and other raw materials being supplied by China in return for basic manufactured and industrial products from Taiwan—could form the basis for growing interdependence between the two sides. Just as it is hoped that the border between Hong Kong and southern China will gradually disappear, so, it is hoped, will the obstacles that currently separate Taiwan from the China mainland.

While this attractive scenario provides what Greenwood and others have called "proliferation disincentives," it has two major fallacies. First, it is probably impossible for Taiwan to obtain any greater measure of energy security from linking up with the China mainland precisely because the PRC itself faces massive energy deficits and will likely do so for the rest of the century. It is clear, for example, that inadequate energy supplies are a major constraint on Chinese economic growth. Second—perhaps even more important—this scenario ignores the role of politics. Above all, the Taiwan problem is a question of politics, involving succession matters in both capitals as well as complex power relationships in each society. It is not merely a difference of opinion between the Nationalists and the Communists that has divided China for thirty-plus years; at stake are a plurality of interests, the most important involving the people of Taiwan. Unless both Beijing and the current government in Taiwan can satisfy the political concerns of the Taiwanese inhabitants, economic interdependence will not lead to political interdependence—unless coercive means are applied.

Time, of course, is essential for the post–Deng Xiaoping and Chiang Ching-kuo successions to work themselves out and determine whether the PRC will stay on its current pragmatic course. There is no doubt that the expertise accumulated by Taiwan over the last three decades could be put to good use on the China mainland in support of the so-called four modernizations, especially if Taiwan's attempt to develop high-technology industry is successful. On the other hand, continued economic success on Taiwan may be the biggest obstacle to a settlement with Beijing. Whether or not that willingness to cooperate will be forthcoming remains a political question that may not be answered for a long time. The main worry is that the longer it takes to answer this question, the more unlikely it is that the answer will be to Beijing's satisfaction. This is a sobering thought in view of the prevailing policy in the West, which is to encourage both sides to take their time.

NOTES

Special thanks to Ms. Karla Brom for research assistance in preparing this paper.

1. For an overview of Taiwan's economic performance, see Shirley W. Y. Kuo, *The Taiwan Economy in Transition* (Boulder: Westview Press, 1983).

2. Energy Committee, Ministry of Economic Affairs, *The Energy Situation in Taiwan, Republic of China*, Taipei, April 1983.

3. These numbers represent *probable* demand. The *lower* estimates are 73,825 million kwh for 1990 and 129,124 million kwh for 2000; the *higher* estimates are 87,533 million kwh for 1990 and 234,365 million kwh for 2000. See Po-chih Lee, "Power Sources Planning Under Conditions of Uncertainty in the Republic of China in Taiwan," *Energy Quarterly* (Taipei), (July 1988): pp. 114-36.

4. Lee, "Power Sources Planning," pp. 116-17.

5. See the article by Thomas Gold in Edwin Winckler and Susan Greenhalgh, eds., *Contending Approaches to the Political Economy of Taiwan* (Armonk: M.E. Sharpe, 1988).

6. Neil Jacoby, *U.S. Aid to Taiwan: A Study of Foreign Aid, Self-Help and Development* (New York: Praeger, 1966).

7. See Denis Fred Simon, *Taiwan, Technology Transfer and Transnationalism: The Political Management of Dependency* (Boulder: Westview Press, forthcoming).

8. See Denis Fred Simon and Chi Schive, "Taiwan's Informatics Industry," in F. Rushing and C. Brown, eds., *National Policies for Developing High Technology Industries* (Boulder: Westview Press, 1986). See also *Japan Economic Journal* (Nov. 15, 1983): p. 5.

9. Government Information Office, *Yearbook of the Republic of China, 1988*, Taipei, 1988, p. 219.

10. *Asian Wall Street Journal*, Oct. 27, 1981, p. 1.

11. For a discussion of Taiwan's concerns about OPEC oil price rises in the early 1980s, see *Journal of Commerce*, March 2, 1981, p. 18.

12. For example, when OPEC decided to cut oil production in early 1987, Saudi Arabia informed Taipei that it would reduce oil supplies to the island by 20 percent. *Free China Journal* (January 26, 1987): p. 4.

13. Energy Committee, Ministry of Economic Affairs, *Energy Policy for the Taiwan Area*, Taipei, 1973.

14. "Pursuing the Nuclear Option," *Far Eastern Economic Review*, July 25, 1980, pp. 44-45.

15. Selig Harrison, *China, Oil and Asia: Conflict Ahead?* (New York: Columbia University Press, 1977). See also Marwyn Samuels, *Contest for the South China Sea*, (New York: Methuen Publishers, 1982).

16. *Free China Journal*, Oct. 20, 1985, p. 4.

17. Premier Yu Kuo-hwa emphasized further diversification as a result of the escalating insurance premiums of U.S.$1 million per month that must be paid for oil deliveries from the Middle East. *Free China Journal*, June 24, 1984, p. 2. For example, in 1984, Taiwan signed an oil import agreement with Ecuador to purchase 10,000 barrels a day; however, this was less than 3.0 percent of Taiwan's daily oil imports. Energy Committee, Ministry of Economic Affairs, August 1984.

18. When energy efficiency did not improve and conservation measures were not adopted by industry as well as residential users in the aftermath of the OPEC crisis, the government and CPC eliminated subsidies and allowed the domestic price of oil to rise to help reduce usage and encourage conservation. See also Ministry of Economic Affairs, Energy Committee, *Energy Management Law* (Taipei, 1980).

19. Kuo Ko-hsing, "An Economic Analysis of Nuclear Power in Taiwan," *Energy Quarterly* (Taipei) 14, no. 1 (January 1, 1984): pp. 143-46.

20. For example, see excerpts on energy policies and development in *The ROC's Four Year Plan for Economic Development of Taiwan, 1982-1985*, Council for Economic Planning and Development, Executive Yuan, Taipei.

21. *Asian Wall Street Journal Weekly*, March 11, 1985, p. 12.

22. Chen Sun and Lian Ci-yuan, "Energy Policies of the ROC, ROK, and Japan: A Comparison," *Industry of Free China*, September 25, 1980, pp. 2-16.

23. See, for example, A. Doak Barnett, *U.S. Arms Sales: The China-Taiwan Tangle* (Washington, D.C.: Brookings Institution, 1982).

24. See, for example, "Are Saudis Switching?" *Free China Journal*, October 20, 1988, p. 1.

25. Richard Solomon, ed., *The China Factor: Sino-American Relations and the Global Scene* (Englewood Cliffs, N.J.: Prentice Hall, 1981).

26. See Foreign Broadcast Information Service (FBIS), *People's Republic of China*, May 8, 1985, p. V2.

27. Chi Hsin, "The Impact of the Hong Kong Accord on Taiwan," *Chi-shih Nien-tai* (The Nineties), no. 11 (Nov. 1984): pp. 51-55.

28. The essence of the "one country, two systems" formula is that Beijing will allow various entities to retain their current modes of political and economic organization in return for acceptance of PRC sovereignty. Ironically, the concept was first articulated in Taiwan in the late 1970s, but was soon after discarded.

29. In August 1988, Taiwan authorities lifted the ban on raw material imports from China. They issued a list of twenty items, including coal, pig iron, and tin, that can be imported. See "New Policy Allows Mainland Raw Material Imports," *Kyodo*, August 4, 1988, translated in *FBIS-CHI-88-151*, August 5, 1988, p. 56. For a detailed analysis of PRC energy resources, see Kim Woodard, *The International Energy Relations of China* (Stanford: Stanford University Press, 1980).

30. *China Post*, April 12, 1985, p. 12.

31. *Hong Kong Standard*, May 14, 1985, p. 3.

32. C.K. Yen, "Energy Economics and Policies," *Industry of Free China*, (November 1982): pp. 1-9.

33. S.L. Chu et al. "Completion of Taipower's First Nuclear Power Unit Through Coordinated Efforts," in American Nuclear Society, *Transactions of the Second Pacific Basin Conference on Nuclear Power Plant Construction, Operation and Development*, Sept. 25-29, 1978, Tokyo, Japan.

34. Y.Y. Pai, "Preliminary Study of Taiwan's Energy Economy," *Industry of Free China* 34, no. 1 (July 25, 1970): pp. 2-25.

35. *Asian Wall Street Journal Weekly*, March 11, 1985, p. 12.

36. See *Free China Journal*, March 9, 1987, p. 1. See also *Taiwan—Country Profile, 1987-88*, comp. *The Economist*, (London, 1988), p. 20.

37. As of the end of 1987, Taipower maintained its commitment to build a fourth nuclear generating plant to come on stream in 1996. *Free China Journal*, December 21, 1987, p. 4.

38. *Asian Wall Street Journal*, October 8, 1981, p. 3.

39. Taiwan energy officials have been eager to join hands with U.S. firms in the construction and design of nuclear power plants in third countries. See Simon, *Taiwan, Technology Transfer and Transnationalism*, chap. 7.

40. There are rumors that Taiwan may be buying uranium from Bolivia, Paraguay, and Canada. Taipower has also signed a cooperation agreement with the Rocky Mountain Energy Company of the U.S. for joint exploration of uranium in seven states. *Free China Journal*, Oct. 20, 1095, p. 4, and March 24, 1986, p. 4.

41. FBIS, *People's Republic of China*, July 7, 1982, p. V4.

42. See Ralph Clough, *Island China* (Cambridge: Harvard University Press, 1978), pp. 116-20. See also *Washington Post*, August 29, 1976, p. 1.

43. See Joseph Yaeger, "Taiwan," draft paper, Feb. 28, 1979.

44. *Free China Journal*, September 8, 1985, p. 4.

45. FBIS, *People's Republic of China*, March 3, 1981, p. B3.

46. *Free China Journal*, May 22, 1983, p. 4.

47. Taiwan researchers have successfully developed a new technology for making nuclear fuel pellets of uranium dioxide by compressing ammonium uranium carbonate into fuel pellets. *China Post*, April 2, 1982, p. 10. Later in 1982, the Institute of Nuclear Energy Research developed a complementary process to extract titanium oxide from heavy sand with an aim towards producing yellowcake. FBIS, *People's Republic of China*, October 7, 1982, V4.

48. *Journal of Commerce*, October 30, 1981, p. 19. A number of questions have also arisen regarding the long-term coal supply contracts that Taipower signed with the United States in the early 1980s. The main issue seems to be why Taipower went ahead with these contracts at seemingly unfavorable prices when its other sources of supply— Australia and South Africa—were seemingly offering better rates. See Economist Intelligence Unit, *Taiwan*, EIU Country Report no. 3, London, 1988, p. 18.

49. FBIS, *Asia-Pacific*, January 22, 1981, p. B4.

50. *Taiwan Statistical Data Book* (Taiwan: Council for Economic Planning and Development, 1987), p. 109.

51. *Nuclear Engineering International*, November 1982, pp. 45-47.

52. Allen Whiting, "China's Foreign Policy Options and Prospects: Towards the 1990s," in Samuel Kim, ed., *China and the World: Chinese Foreign Policy in the Post-Mao Era* (Boulder: Westview Press, 1984), pp. 321-334.

53. Chalmers Johnson, "The Mousetrapping of Hong Kong: A Game in Which Nobody Wins," *Asian Survey* 24, no. 9 (September 1984): pp. 887-909.

54. *Beijing Review* 24, no. 40 (October 5, 1981): p. 10.

55. Chiu Hungdah, "The China-Taiwan Unification Question: An Analysis of the View of the Republic of China," *Asian Survey* 23, no. 10 (October 1983): pp. 1081-1094.

56. See Thomas Gold, *State and Society in the Taiwan Miracle* (Armonk: M.E. Sharpe, 1986). See also Walter Galenson, *Economic Growth and Structural Change in Taiwan* (Ithaca: Cornell University Press, 1979). On the PRC side, see Harry Harding, *China's Second Revolution: Reform After Mao* (Washington, D.C.: Brookings Institution, 1988). See also Cheng Chu-yuan, *China's Economic Development: Growth and Structural Change* (Boulder: Westview Press, 1982).

57. Ramon Meyers, "The Contest Between the Two Chinese States," *Asian Survey* 23, no. 4 (April 1983): pp. 536-52. See also K.T. Li, *The Evolution of Policy Behind Taiwan's Development Success* (New Haven: Yale University Press, 1988).

58. For a discussion of recent changes in the political system, see John Cooper, *Taiwan's Elections: Democratization and Political Development in the Republic of China* (Baltimore: University of Maryland School of Law, 1984).

59. "Question of Taiwan Independence," *Asian Wall Street Journal Weekly*, August 15, 1986, p. 16.

60. John Quansheng Zhao, "An Analysis of Unification: The PRC Perspective," *Asian Survey* 23, no. 10 (October 1983), pp. 1095-1114.

61. "China Says U.S. is Trying to Foil Effort on Taiwan," *Boston Globe*, Nov. 6, 1988, p. 11.

62. See Martin Lasater, *The Taiwan Issue in Sino-American Strategic Relations* (Boulder: Westview Press, 1984).

63. See Lasater, *The Taiwan Issue*.

64. Ibid.

65. For example, see FBIS, *People's Republic of China*, June 17, 1982, p. V2, and December 9, 1983, p. V1.

66. Edwin Synder et al., *The Taiwan Relations Act and the Defense of the ROC* (Berkeley: University of California, 1980).

67. FBIS, *People's Republic of China*, June 3, 1985, pp. W1-34.

68. Ibid., October 17, 1984, p. V1-2.

69. Gerald Segal and William Tow, *Chinese Defense Policy*, (Champaign/Urbana: University of Illinois Press, 1985).

70. "Mainland Invites Taiwan Investors," *Beijing Review*, July 18, 1988, p. 9.

71. "Nuclear Scientist Said to Admit Spying for U.S.," *CNA-Taipei*, June 15, 1988 translated in *FBIS-CHI-88-116*, June 16, 1988, p. 65.

72. In mid-1985, the Hsinhsin Research Center was set up by the government "to conduct basic leading research related to weaponry of the future." FBIS, *People's Republic of China*, May 8, 1985, p. V1.

73. Taiwan's defense budget in 1985 was NT$141.8 billion, accounting for 39.4 percent of the total expenditures of the central government. FBIS, *China Political, Military and Sociological Affairs* (CPS-84-030), April 20, 1984, pp. 71-72.

74. *Aviation Week and Space Technology*, April 12, 1982, pp. 46-50. For example, Taiwan has been anxious to develop its own jet fighter to replace its U.S.-designed, aging F5E. Both the Chungshan Institute and the Center for Aeronautics Research have begun design of an aircraft known as the XF. FBIS, *People's Republic of China*, January 17, 1982, p. V2. Similarly, in May 1985 Taiwan developed its first ground-to-air missile, which is similar to the U.S.-designed Patriot. FBIS, *People's Republic of China*, May 8, 1985, p. V1.

75. In this regard, a multi-ministry "national defense technology committee" was established in 1985 to accelerate weapons development. FBIS, *People's Republic of China*, March 4, 1985, p. V2.

76. James Hsiung, ed., *Beyond China's Independent Foreign Policy: Challenge for the U.S. and Its Allies* (New York: Praeger, 1985).

77. FBIS, *People's Republic of China*, October 3, 1983, p. V1-2.

78. Ibid., January 18, 1985, p. V2-3.

79. Samuel Kim, "China and the Third World," in Samuel Kim, ed., *China and the World*, pp. 178-214.

80. "Arab Lands Said to Be Turning to China for Arms," *New York Times*, June 24, 1988, p. A3.

81. See "A Survey of Taiwan," *The Economist*, March 5, 1988, p. 13.

82. See *Free China Journal*, July 7, l986, p. 1.

6

Argentina

CYNTHIA A. WATSON

Argentina has baffled both internal and external observers for nine decades of the twentieth century with behavior that has not seemed reasonable when compared with the opportunities available to the people. Nationalism, the desire to control Argentina's growth and destiny, is a strong force in the republic. While a highly controversial figure, Juan Domingo Perón described the feelings of many Argentines when he said: "This is in a few words the Argentine international doctrine: we want to respect all peoples but we want to be respected in turn by them; we are always on the side of the subjugated because we understand that in the community of peoples of the world there cannot be powerful people who possess everything while there are weak people who suffer everything."[1]

Regardless of the reasons for the difficulty in explaining or predicting Argentina's political and international behavior, the nation at the southern tip of Latin America is a threshold nuclear weapons state. The nuclear program has been cultivated carefully and is a source of significant national pride. It gives the republic some options in addressing both its energy needs and its military strategies. This chapter will consider these options within the context of the overall energy picture.

International concern about Argentina's nuclear program heightened after the south Atlantic conflict with Britain in 1982. The world press highlighted Argentine capabilities in astonishment that this somewhat peripheral state would take on the power and prestige of the Royal Navy. The majority of the world was not aware of the conflict that had simmered between Washington and Buenos Aires during the years immediately following World War II, nor of the national power envisioned by Perón during his decade as leader. Argentina was too far away to be of concern to many nonspecialists, often mentioned only because of the seemingly constant transitions of governments, more reminiscent of a banana republic than an aspiring world player. The horrors of the National Security State, which began in 1976 with the ouster of Isabel Perón, received some attention on the international scene through the efforts of human rights groups and the Carter administration in the United States. The end of the

Carter period appeared to indicate the end of international awareness in 1981, as the United States and other industrialized democracies elected more conservative governments. The newly elected governments seemed more interested in promoting business links than human rights in Argentina.

The stark realization that Argentina was not merely a peripheral state hit both policymakers and the public hard. The nuclear capability should not, however, have come as any surprise. Argentina's aspirations to create a world-class nuclear program date back to a rather humiliating episode in the early 1950s. After Perón announced to the world that he had created the first-ever controlled thermonuclear explosion, only to find that his "scientist," former Nazi Ronald Richter, was a charlatan,[2] Perón set out to create a strong, legitimate nuclear energy program. While Argentina has remained adamant since the 1950s that its goals are entirely peaceful, the global community has been somewhat suspicious.

This chapter will explore the nexus of energy and security in the context of contradictions between nationalism and the need to reach outside its borders that have plagued Argentina. It will discuss the political realities behind these phenomena. In the process, it will focus particular attention on Buenos Aires's relatively sophisticated nuclear program.

Energy in Argentina: The Early Years through the 1960s

Argentine petroleum deposits were discovered in the area around Comodoro Rivadavia in December 1907. The initial years of oil development saw Standard Oil of New Jersey, Royal Dutch Shell, and an Argentine company, Astra, working in relatively advantageous conditions, since the Argentine regime feared that its fuel supplies might be cut off by impending conflict in Europe.[3] The problems of World War I mandated a national program for petroleum development. The national petroleum agency, Yacimientos Petroliferos Fiscales (YPF), was created in 1922 for the purpose of producing petroleum and related activities. Although YPF had government support, it was unable to make Argentina self-sufficient through the 1940s.[4] Demand for petroleum rose at a higher rate than did production, particularly at the end of World War II, because of the dramatic rate of industrialization. From 1947 until 1955, petroleum imports climbed 100 percent to meet consumption, as YPF fell short of goals in increasing its domestic operations. The foreign corporations operating under contract with the YPF were prohibited from increasing their production levels, hence closing another method of holding down imports. YPF had not generated enough profits to invest in newer technologies and was allowing Argentina to fall behind its government-controlled petroleum production until 1940 while YPF increased refining of foreign companies' oil.[5]

Perón and his rhetorical concerns about national sovereignty had created domestic expectations that Argentina would be able to produce enough petroleum to meet its needs. By 1953, Perón had given way to economic realities and allowed foreign investment in the petroleum industry, but the damage had already been done. With Perón's ouster in 1955, the national sentiment for sovereignty over domestic resources collided with the need for oil. Arturo Frondizi, a successor to Perón who charged that the Argentine people should retain control over their assests, had given YPF a legal monopoly over the production of petroleum by 1958.[6] However, Argentina then had to buy 7 million barrels of crude oil from the Soviet Union while Frondizi attempted to entice private petroleum firms, some of them foreign, into production in Argentina. Buenos Aires signed drilling and development contracts with many private firms, including Esso, Royal Dutch Shell, Kerr-McGee, Astra Cia. Argentina de Petroleo, and Cia. Argentina para El Desarrollo de la Industria Petroleo y Minerales, SA.[7] The need for increased energy production conflicted with the Argentine desire for national control over the means of production as well as profits. When Frondizi saw that YPF was unequal to the task of raising production, he turned to the private sector in a change of policy that was unpopular but that increased domestic production while drastically decreasing imports.

Beyond oil, the state became involved in other aspects of energy production. The government body Agua y Energía took control over foreign holdings in electricity, while Gas del Estado, established in 1957, oversaw natural gas production.[8] Even with growing state control, it was not until the latter 1960s that Argentines believed themselves on the way to self-sufficiency in various types of energy supplies. As late as 1957, the Frondizi government was accused by the opposition of bringing the republic to the point of importing two-thirds of its petroleum needs.[9] Energy sufficiency was a domestic political football as opposition figures charged that contracts with foreign companies were actually concessions which were costing Argentine money and its natural resources. While the government was working through foreign companies in developing domestic energy, it was creating a parallel Argentine nuclear program aimed increasingly at autonomy. The Frondizi government's attempts to obtain foreign financial, managerial, and technical assistance were exemplified by the so-called Yadarola plan, whereby the Argentine ambassador to Washington tried to raise $1 billion in outside capital for petroleum exploration and production.[10] The Frondizi plans were successful between 1957 and 1959, when Standard Oil, Royal Dutch Shell, and other corporations contracted to aid Argentine energy development. Outside investment was also obtained for petroleum and natural gas pipelines, as well as hydropower expansion.[11]

The Frondizi regime's willingness to engage foreign private participation in the late 1950s, less than a decade after Perón and his rhetoric of nationalism, was to set a pattern of contradictions throughout the next three decades. Argentina's energy policy has alternated between high dependence on the government's role and reliance on private—often foreign—participation. These shifts in reliance on state versus private sources of expertise and funding resulted from political necessities and economic realities during the turbulent years of the 1960s, 1970s, and 1980s. These dramatic shifts between nationalism and foreign assistance have characterized Argentine energy development in the 20th century.

During the Alianza Para el Progreso of 1961-69, Argentina continued to receive energy-related loans from the U.S. government as well as the Inter-American Development Bank, the Export-Import Bank, and the World Bank.[12] Such funding came under the control of the state corporations YPF and Agua y Energía, which then worked to create joint participation where possible. This pattern of joint participation became entrenched as the main option when state operation of industries was jettisoned with changes in government. State enterprises have contracted operations out to foreign companies for lack of technical and financial resources. In the case of petroleum and natural gas, foreign companies often sold specified amounts of production to the state at prices under the world market levels.[13]

The main sources of capital and technology for energy development during the three decades have been Western, whether private or governmental; the Soviet Union had limited participation in Argentina's energy development until 1970, but the two states increased nuclear power ties in the 1980s.[14] The United States, through the Atoms for Peace program, worked to strengthen the Argentine nuclear energy program in the 1950s. The first experimental nuclear plant in Latin America arrived in Argentina in 1957 from Westinghouse—the RA-1 reactor, located 50 kilometers outside of Buenos Aires at the Ezeiza Atomic Center. It resulted from a 1955 agreement under which the United States pledged to supply the reactor while training 200 Argentine scientists and providing $460,000 in grants from the U.S. Atomic Energy Commission. The RA-1 was a light water reactor for which enriched uranium was imported from the United States. The reactor was to become a source of resentment in Argentina when the Carter Administration curtailed sales of enriched uranium in the late 1970s in retaliation for Argentine human rights abuses.[15]

In one of the periods when foreign enterprises' development of Argentine resources clashed with nationalist sentiments at home, Arturo Illia, Radical party president, in 1963 canceled contracts held by foreign corporations without paying compensation.[16] The United States responded by terminating foreign assistance, forcing Illia to pay the foreign oil companies compensation of more than $800 million.[17] By 1965, government policy had reverted to greater Argentine, rather than outside, control over national assets.[18] When the military took power the following year, foreign corporations were invited back to resume development of energy resources, as would happen again in the 1970s.[19] During General Juan Onganía's tenure (1966-70), foreign corporations explored for petroleum and natural gas even *without* supervision of the Argentine state enterprises.[20] This policy came under severe attack under the rule of a Perónist, Alejandro Lanusse (1971-1973), whose government was quickly overturned. During the brief Perónist interlude of the 1970s, full-scale nationalization was threatened but never conducted,[21] although Perón did move all petroleum marketing back under control of YPF.[22] YPF became a state-owned company in 1977, with somewhat greater autonomy, although it remained under the Energy Secretary. YPF, now the largest enterprise in the republic, owns service stations; it is divided into a public section and a military/industrial section.[23]

Petroleum production in Argentina rose to the point of near-independence in liquid fuels by 1982 (Table 6.1). During the same period, natural gas production rose at a fairly consistent pace, from 1,058,000,000 cubic meters in 1955 to 15,523,000,000 cubic meters in 1982.[24]

Development of Nuclear Capabilities: 1950 through 1984

Argentina's nuclear energy agency, the Comisión Nacional de Energía Atomica (CNEA), was set up in the 1950s as a result of international embarrassment over the charlatan Richter. Perón's goal was then to create a solid nuclear organization that would be above reproach (including membership by anti-Perónists who otherwise might not have been considered acceptable).[25] To avoid dependence on states that might use political leverage to subjugate Argentina,[26] the CNEA was created with a highly organized, well-defined mission to help create self-sufficiency in energy. Some thirty years after its establishment by Pedro Iraolagoitia, the Jorge A. Balseiro Institute of Physics at San Carlos de Bariloche remains a respected nuclear training facility. The institute has evolved into a public policy think tank as well as a CNEA education facility. Regardless of the other problems the nation faces, it has retained a substantial pool of technical expertise.

The first nuclear reactor in Argentina was a heavy water–natural ura-

Table 6.1. Argentina's imports of petroleum and its byproducts, 1955-1982 (Thousands of cubic meters except as noted)

	1955	1960	1965	1970	1971	1972
Petroleum	4,621	3,685	4,203	1,684	2,540	1,736
Aircraft	66	64	85	—	—	—
Kerosene	211	596	16	—	62	—
Gasoil	465	589	758	627	239	—
Diesel oil	195	316	55	9	6	—
Fuel oil	2,713	309	26	168	148	57
Propane*	—	—	—	103	154	91
Butane*	—	—	—	273	337	266
LNG*	—	26	183	376	491	357
Lubricants	80	63	107	142	172	54

	1973	1974	1975	1976	1977	1978
Petroleum	3,395	3,430	2,486	3,524	3,413	2,476
Aircraft	—	12	14	5	5	—
Kerosene	42	—	60	—	—	34
Gasoil	107	114	550	500	590	179
Diesel oil	—	20	—	—	—	—
Fuel oil	436	80	—	99	—	—
Propane*	69	141	115	126	62	114
Butane*	237	271	305	249	171	260
LNG*	306	412	420	375	233	374
Lubricants	1	3	39	43	23	29

	1979	1980	1981	1982
Petroleum	2,009	2,529	1,447	823
Aircraft	—	—	—	—
Kerosene	249	100	117	—
Gasoil	969	15	19	—
Diesel oil	—	—	—	—
Fuel oil	—	—	—	—
Propane*	194	77	78	—
Butane*	231	232	154	—
LNG*	425	309	232	—
Lubricants	13	23	28	19

Source: Anuario Estadistico del la Publica Argentina: 1981-1982, Instituto Nacional de Estadistica y Censos, Buenos Aires, 1982, p. 531.
*Thousands of tons.

nium type sold by the West German conglomerate Siemens, which went on stream in 1974. As early as 1964, President Illia had requested that the CNEA conduct a feasibility study for nuclear power generation in the Buenos Aires metropolitan area, which resulted in a decision to seek bids on a 500-megawatt reactor to supply electricity by 1971. The study also set two benchmark goals: to utilize as much as possible of the abundant indigenous uranium supply as possible, and to accomplish roughly 40-50 percent of the construction and operation of the power station through domestic participation.[27] This study was one of the earliest accomplished by a state independently, rather than through Atoms for Peace or the International Atomic Energy Agency.[28]

Among the 17 bids from the United States, West Germany, France, Canada, and Britain, both the Canadians and West Germans offered reactors based on the heavy water–natural uranium model, which fulfilled the CNEA feasibility study's goal of high utilization of domestic uranium supplies without exporting fuel for enrichment. CNEA accepted the Siemens bid for a 320-Mw reactor although it was not the lowest-priced, nor had Siemens even tested a reactor of 320 Mw.[29] However, Siemens offered Argentina 35 percent local participation and 100 percent financing. Members of the Buenos Aires government also had personal links with Walter Schnurr, a West German who lobbied for Siemens. Since the CNEA was convinced that the West German proposal would ultimately increase Argentine autonomy, the $70 million contract was signed by supporting local participation, which could translate into nuclear independence. Construction on Atucha I began at a site 150 kilometers north of Buenos Aires on the Paraná River. While scheduled for completion in 1971, Atucha I did not go critical until March 1974, because of a variety of problems. Siemens absorbed the expense of redesigning the problematic fuel rods, as well as cost overruns, which may have been 100 percent.[30] The government in Buenos Aires was impressed with the West German willingness to satisfy the customer, which influenced later power plant decisions and solidified existing nuclear links. This connection generated some international concern when stories surfaced that the West Germany, which is prohibited from developing nuclear weapons, may have secretly provided illegal nuclear materials ("hot cells") to Argentina. These stories circulated widely during the 1982 conflict between Britain and Argentina but have not been pursued since.[31]

Atucha I proved successful. In industry efficiency reports, it was rated first of 154 reactors in the world between 1977 and 1978.[32]

In 1967, CNEA had developed ten-year plan for nuclear development, setting goals of three power plants, domestic fuel manufacturing and disposal, and up-to-date breeder and plutonium technology. Partially as a result of this report, Argentina had five research reactors by 1971 and a small domestically designed, produced, and operated reprocessing facility (the first in Latin America for plutonium).[33] When the 1967 nuclear plan was completed, Rio Tercero Lake in Córdoba province in the interior was chosen as the site of the next reactor, the Embalse plant. The contract for the 600-Mw heavy water–natural uranium facility went to the Atomic Energy of Canada, Ltd. (AECL) of Canada, after considerably greater public debate than in the late 1960s, when Atucha I was awarded. The Canadians did not offer 100 percent financing for Embalse, but their proposal was appealing in calling for transfer of additional nuclear technology.[34] Skeptics of the Argentine program pointed out that the Atomic Energy of Canada CANDU type of heavy water reactors produced greater amounts of plutonium than other reactors, a characteristic that might have been the deciding factor if the republic were seeking to acquire nuclear weapons.[35] The CNEA got the AECL to agree to 50 percent Argentine participation[36] for the $250 million project. The Embalse reactor contract was signed in 1971, but construction was delayed for two years—only the beginning of the reactor's problems.

The Embalse reactor became a classic example of the bureaucratic/ economic/international problems facing Argentina as a society and a threshold state. The year before construction began, Juan Domingo Perón returned from a seventeen-year exile and a period of political collapse began, culminating in the National Security State and the *guerra sucia*. While Perón had carefully avoided politicizing the CNEA after the Richter fiasco in the 1950s, during the Perónist Interlude of the 1970s it became a highly politicized organization. Decision-making disintegrated because of political purges and the lack of direct communication between the CNEA and the Casa Rosada (the head of state works out of this site).[37]

More significant, the Canadians felt betrayed by the Indian nuclear detonation of May 18, 1974, which proved that a state could develop a nuclear explosive from a civilian power plant. The Canadians felt at least partially responsible for the Indian action, as they had supplied much of the technology for the Indian program, which began in the 1950s. After the detonation, Canada attempted to increase the safeguards on the Embalse facility, which was then under construction. The Argentines resisted such attempts, and considerable wrangling slowed construction. If this were not enough, the peso devaluations and inflation spiral of the Perónist Interlude forced the AECL to take $200 million in losses, as the ultimate price tag reached four times what had been contracted in 1971.[38] Embalse went critical in March 1983, leaving serious questions about the nuclear industry's external ties, as well as international concerns about Argentina's goals.

After the collapse of Isabel Perón's government in March 1976, Ret. Rear Admiral Carlos Castro Madero, trained at Westinghouse in the United States and holding a Ph.D. in physics, assumed charge of the CNEA early in the Videla government. He managed to survive all the upheavals that plagued Argentine society and government over the next seven and a half years, remaining in control of the CNEA until January 1984. Under his direction, the Argentine nuclear capability came of age. Madero left the CNEA a viable agency for nuclear development, potentially a technology exporter of the Third World.

The CNEA moved to put its 1967 growth plan more thoroughly into effect with Argentina's third nuclear generating plant. During the Perón period, the CNEA under Iraolagoitia had discussed a third nuclear reactor of the same size as the 600-Mw Embalse, but the tensions with AECL left doubt regarding any Canadian bids. The Siemens delays of the 1960s, extensive though they were, were rapidly fading from memory, and the West German willingness to buck the Ford Administration on supplying a complete nuclear fuel cycle to Brazil[39] made the West German option increasingly appealing again. Building the third reactor at the same location as Atucha I was expected to cut costs, and local energy demands were bound to grow, since one of every three Argentines live in the Buenos Aires metropolitan area. Twin reactors at Atcha made economic and logistical sense. In November 1978, the junta in Buenos Aires appointed an interministerial commission to continue developing a national plan for nuclear growth and to start taking bids for the new reactor.

The CNEA's 1979 plan was to chart nuclear development through the turn of the century. Argentina's plan "is aimed at stepping up the scientific, technical, and industrial effort in the nuclear field for peaceful uses and utilization of the country's human and natural resources in this field to help consolidate national growth".[40] The plan, approved through decree 302/79, charged the nation to develop reactors to go on line in 1987 (Atucha II), 1991, 1994, and 1997. The plan called for a total of 15 reactors (for research as well as power generation) by the year 2000, with generating capacity of 9000 Mw, at a projected construction cost of $10 billion.[41] Decree 302/79 cemented the commitment to heavy water reactors as the sole means of Argentine nuclear power generation. For future facilities, Argentina would be self-sufficient in the heavy water part of the fuel cycle. By the year 2100, the nuclear industry was projected to fill Argentine energy requirements beyond the 93 percent petroleum self-sufficiency to be achieved by then. In 1979, petroleum provided 63 percent of the nation's energy, gas 23 percent, hydropower 6 percent, and nuclear-generated power merely 2 percent. By the turn of the century, Argentina hopes to increase the nuclear contribution to 15 percent, and hydropower to 73 percent while lowering fossil fuels to 12 percent.[42] The full contingent of reactors by 2000 were to be using 600 tons of uranium per year of the known reserves of 30,000 tons in Argentina.

One of the main goals of the plan, as has been true throughout nuclear development in Argentina, was to keep increasing domestic participation in the management and ownership of the nuclear industry. As a result, CNEA moved to create four enterprises as mixtures of public-government participation, as well as domestic-foreign, to get the financing to best develop Argentina's nuclear capabilities. These enterprises included one to help develop nuclear potential in Río Negro and Patagonia provinces and, a second enterprise, in Mendoza, as a mixed mining group in the "province of uranium." The third company, with the majority privately owned and the minority held by CNEA, is responsible for technical production and development of the combustible aspects of irradiation process. Fourth is a mixed engineering enterprise held 75 percent by CNEA, in charge of deciding the reactor sites for the projected three additional reactors.

The Atucha II reactor resurrected international worries about Argentina's elevation to the status of threshold nuclear state. Global nuclear weapons proliferation concerns greatly increased after the 1974 Indian detonation and the Bonn-Brasilia fuel cycle pact of the following year. Not only the Canadians were worried about the possibility that Argentina might derive weapons from the civilian power generating plants; the United States and several European countries were equally apprehensive.

The West Germans, however, were significantly less concerned, since Argentina was already developing many of the nuclear fuel cycle components itself. The West Germans won the contract for the Atucha II reactor, and a Swiss company simultaneously got the contract to build a heavy water production facility. The West Germans clearly wanted the reactor contract enough that they did not care about not getting the heavy water facility as part of the package. While the Carter and Trudeau administrations encouraged the Germans not to go through with the contract, the Atucha II process "pointed out three long-standing problems on this issue . . . 1) the bankrupt state of the Carter policies on nuclear commerce, 2) the sizable number of orders for West German nuclear products, and 3) the military-commercial balance of power in Latin America".[43] Coupled with the emerging information about the brutality of the National Security State in control of Argentina the prospect of Atucha II was unsettling for those concerned about Argentina's status as a threshold state.

Atucha II, a 750-Mw heavy water–natural uranium reactor, was contracted to West Germany's Kraftwerk Union (KWU), while the 250-ton-per-year heavy water production plant at Arroyitos was awarded to the Sulzer Brothers of Switzerland. The heavy water plant, the third in Argentina, consolidated its move toward self-sufficiency. The awards had vocal critics at home and abroad, particularly since other bids were as much as $500 million lower than the winning KWU contract.[44] Critics felt that Argentina, in selecting the reactor, had been more concerned about the few nonproliferation strings attached than about the economics of the system. The CANDU reactors sold by Canadian corporations were more highly tested than the prototype that KWU had offered.[45] The Canadians were reluctant to push their case too strongly, however, since the superb Atucha I reactor also had been a prototype. Nor was the Sulzer Brothers heavy water facility exempt from criticism, since that firm's largest plant to date had been only one-tenth the size of the one for Argentina, as well as being prone to problems.[46] Much of the criticism about the two 1979 contracts centered on whether the CNEA had subdivided the business to circumvent stiff Canadian nonproliferation safeguards.[47] The United States pressured the West Germans and Swiss on grounds that Argentina had not agreed to the mandated blanket safeguards that these suppliers had accepted as members of the London Supplier Club.[48] However, neither the West Germans nor the Swiss abandoned the $1.6 billion combination German-Swiss deal, although in 1980 Bonn got pledges on stiffer safeguards. Castro Madero's stated position on the safeguards:

> Argentina maintains that the international exchange of nuclear technology must be covered by suitable safeguards, as a means of promoting the development of nuclear energy for peaceful purposes. On this basis, it has put under safeguards all the nuclear facilities that it has received from abroad, but it refuses to subject to safeguards those facilities which have been developed and built with its own technology, and wherein it has not received any assistance from abroad. This is the only way of actually giving an incentive to external assistance in areas that have been unilaterally defined as sensitive by the London Club. It refuses to give a blank check, such as the signing of total safeguards, which would leave the country without the capacity to negotiate for the procurement of major nuclear technology. In short, safeguards are accepted in exchange for technology.[49]

The progress on both Atucha II and Arroyitos has been plagued by economic and technical slowdowns, but nuclear energy is helping to meet Argentina's needs.

ALFONSIN AND THE DEBT CRUNCH: STALL ON ENERGY

After the ousting of the leaders of the *guerra sucia*,[50] when Raul Alfonsín assumed office on December 10, 1983, a somewhat different path was

Table 6.2.   Energy production in Argentina, 1977-1987

	1977	1980	1984	1986	1987
Petroleum, 000 bbl/day	431	491	478	434	428
Natural gas, 000 bbl/day	11	13	8	31	31
Dry natural gas, trillion cubic ft.	0.28	0.28	0.49	0.55	0.54
Hydroelectric power, billion kwh	5.7	15.0	19.7	20.8	21.0
Nuclear electric power, billion kwh	1.6	2.2	4.3	5.4	6.1

Source: Office of Energy Markets and End Use, U.S. Department of Energy, *International Energy Outlook, 1989*, Washington, D.C., 1989.

chosen for the nexus between the state and the development of energy in Argentina. Alfonsín did not choose to throw open the country to foreigners nor did he unceremoniously oust them, although public opinion during the 1980s has been high in opposing repayment of the massive external debt accumulated during the 1976-83 period. Alfonsín was forced by the reality of Argentina's debts to make overtures to foreign energy companies, limiting the operations of primarily state-owned energy companies. Alfonsín needed to satisfy the criteria set forth by the International Monetary Fund as conditions for renegotiating its debt servicing schedules. Under its economically orthodox policies, the IMF set conditions for short-term-loans and debt servicing that would limit state control over the economy, including energy enterprises. This led the Alfonsín government to solicit cooperative agreements in energy matters with Italy, Japan, West Germany, and Brazil. The Italian state oil company, Ente Nazionale de Idrocarburi (ENI), agreed on a $350 million joint venture with YPF and Gas del Estado for the annual production of 285,000 tons of LPG. ENI agreed to provide one-third of the financing at 1.75 percent interest for the duration of the 20-year loan.[51] Through the World Bank and the Japanese Export-Import Bank, YPF would receive extra financing for petroleum refining and natural gas development.[52] A Franco-German-Argentine consortium, Total-Deminiex-Bridas, discovered natural gas under San Sebastián in Tierra del Fuego in early 1988. This discovery was expected to raise natural gas reserves to roughly 900,000 cubic meters at total drilling costs of $511 million.[53]

Part of the changes that Alfonsín sought concerned control over the energy agencies, particularly the CNEA. From its inception in the early 1950s through January 1984, the CNEA was under the control of retired navy men. Some of them were engineers chosen for their technical knowledge (such as Madero, the U.S.-trained physicist in control between 1976 and 1984) or for their political affiliations (including Iraolagoitia, a staunch Perónist in charge several times when Perónists took control of the nation between the 1950s and 1970s). Alfonsín realized that much international concern focused on military control of the CNEA and the implied threat that the avowedly civilian nuclear energy program could be used for military purposes, particularly after the southern Atlantic conflict of 1982. He removed Madero on January 24, 1984, replacing him with the civilian engineer Alberto Constantini. Alfonsín also set about to cut the CNEA budget drastically, partially because of the reality of Argentine economic problems and partially the fears aroused by the 1982 conflict, when Madero had threatened to embark on a program to build nuclear-powered submarines for "self-defenses."[54] While terminating military leadership, Alfonsín did not eliminate the long-term navy influence on the program. The government tried to reorganize the CNEA itself by putting the two operational nuclear power plants under the control of a new corporation, but the nationalist opposition feared that this would diminish the nuclear program.[55] Alfonsín's moves to contain the state's role in energy extended to the military ownership of chemical factories in 1987; Petroquímica Bahía Blanca and Petroquímica General Mosconi reverted from indirect military control to that of the civilian-based Department of Defense.[56] Many political opponents in Argentina feared that under Alfonsín the natural resource muscles of Argentina would atrophy in order to satisfy the foreign creditors, their concerns heightened by the president's desire to expand the energy sector during the last years of the century. As long as Alfonsín's austerity programs retained fairly broadly based political support, he could continue his programs, but the periodic devaluations of the austral and the increasingly severe economic belt-tightening reduced his flexibility in terms of Argentine energy and political will.

Expansion plans for the energy sector appeared in 1985 when the vice-minister for energy planning unveiled a program that reserved 55 percent of all public investment for energy development, particularly petroleum and hydroelectric ventures. Two additional nuclear power plants were anticipated, as under the 1979 nuclear plan.[57]

The target for investment during 1985 was $3 billion over a fifteen-year period, but the feasibility of such high levels of investment was questionable, particularly when coupled with the perennial renegotiations of the debt repayment schedule with multilateral lending organizations. By the end of 1985, the economy minister reported that the International Bank for Reconstruction and Development and the Inter-American Development Bank had credited Argentina with $300 million for energy-related projects,[58] for which Alfonsín's government had appealed across the international community. However, the participation by international corporations has fueled concern about Argentina's losing its sovereignty to outsiders in selling its vast natural resources. The "Olivos plan" for selling petroleum illustrates the paradox. The plan was an acknowledged improvement over the military's outright sale of subsoil rights to foreigners, but the government's agreement to pay no less than 80 percent of the world petroleum price, in hard currency, to independent suppliers nevertheless aroused some concerns, since it meant diminishing foreign currency reserves.[59]

Alfonsín did not alter the development of nuclear energy as dramatically as other energy sectors, for several reasons. For one, the CNEA has been successful at operating the nuclear energy program without outside interference, as noted, due to its own structure and the nature of the industry itself. Once Argentina began developing reactors, deliberate steps were taken to decrease external influence on the program. A second differentiation has been the strong national sentiment evoked by the nuclear energy program's success. Many Argentines who supported the president in his attempts to put the nation back on a more stable path were angered at the idea that Alfonsín might destroy the perceived success of this autonomous program. While acknowledging that Argentina had bought the Embalse and Atucha I and II reactors from other nations, Argentines cling to the goal of domestic production of nuclear reactors and to the ambiguous status of being a nuclear force. Upon taking office, Alfonsín and Foreign Secretary Dante Caputo indicated that they were considering making Argentina a full party to the 1967 Treaty of Tlatelolco. The Perónist party was joined by people across the political spectrum in claiming that Argentina's sovereignty was more imporant than placating the international community on nuclear safeguards. Since it procured its initial reactor in 1968, Argentina has steadfastly declared that it has only peaceful goals for nuclear power but that it cannot allow the industrialized nations to dictate Argentine domestic policy. Argentina was sovereign not only in its development of nuclear power but also in its evolving role as nuclear supplier; Tlatelolco would have restricted that activity. Scientists also feared that one of the Third World's most vibrant research programs would be hurt. Alfonsín's reconsideration of the Treaty of Tlatelolco, coinciding with realization of the $50 billion external debt, appeared to show Argentina as giving in to foreigners at every turn. After strong lobbying by many interested parties, the major political parties issued a joint statement on nuclear development and research, saying that it "is a fundamental pillar of out national liberation and growth."[60] Overall energy production in Argentina was not consistent, however, as shown by massive power shortages in late 1988 and 1989.[61]

Although Alfonsín's moves created a domestic uproar, he was only reacting to international apprehension about the country's nuclear program. With Castro Madero's announcement—coming between Alfonsín's

election and his accession to power—that the CNEA had built a clandestine uranium enrichment facility, questions regarding Argentine intentions multiplied. During the 1982 conflict, Castro Madero's comments
about a nuclear-powered submarine, coupled with Buenos Aires's stub-
bornness on signing international nonproliferation agreements and the
revelations about uranium enrichment, made nuclear proliferation specialists skeptical that Argentina was being honest about its goals. Frantic
Argentine declarations that its sovereignty should not be violated coin-
cided with requests by Buenos Aires that the economic community make
concessions in debt repayment. Alfonsín's position was complicated by the
October 1986 statement by the chief naval officer that Argentina's first
nuclear submarine (with three more planned) would sail in two years "if
budgetary restrictions did not slow us down".[62]

This statement was a further indication of domestic pressures to pursue
the nuclear program in spite of budgetary problems and international
pressures to curtail the program, putting Alfonsín in a no-win situation. He
had reduced the budgets of the CNEA and all armed services by significant amounts in the early months of his administration, yet indications
were that the public desired continuation of these projects. At issue in such
pressures was the same question that Alfonsín and his successors will face
for the foreseeable future: How will Argentina juggle debt repayment with
domestic spending projects believed essential to the national security, be
they nuclear energy or more conventional forms? The issue is not merely
whether the government or the public has control over the energy sector;
rather, it is a priority struggle between the domestic goals and international demands, both of which have tremendous clout with the person in
the Casa Rosada.

While Brazil is increasingly considered an international economic
giant, with the world's ninth largest economy, and its economic and
political power creates new challenges for Argentina. The concept of
subregional integration of Brazil and Argentina took off in 1985 after the
elected governments of the two came into office. Given their history of
geopolitical suspicion, critics were somewhat skeptical of the ideas of
economic integration put forth by Alfonsín and Brazil's Jose Sarney. Alfonsín was under particular pressure not to enmesh Argentina in any
scheme that would further deprive the nation of its economic autonomy;
critics feared that the massive Brazilian economy would overwhelm Argentina's. While the two presidents met every six months to discuss mutual
concerns and the progress made in integration, the process moved slowly.[63] A major part of the integration was the sharing of nuclear technology
and a lessening of the threat that the two states would allow their civilian
nuclear power programs to become the bases for competition in nuclear

weapons. Argentina's nuclear program, significantly ahead of Brazil's, evoked concern in Brazil when the CNEA Chairman Madero, triumphantly announced that Argentina had developed its own nuclear enrichment program. Observers both inside and outside Argentina and Brazil believe that agreements resulting from the biannual meetings between Alfonsín and Sarney will not only make it more difficult for either state to feel threatened by the other but will also enhance the sharing of appropriate nuclear technology between the two.[64]

GROWING APPREHENSION ABOUT THE NUCLEAR OPTION

Conventional wisdom on nuclear arms has held for thirty years that nonindustrialized states of the Third World would not have significant delivery capabilities for the foreseeable future. That the newer nuclear states would have the delivery capabilities to hit the United States or the European mainland seemed inconceivable, since intercontinental ballistic missiles were the purview of the established nuclear powers, outside the control of those newly emerging. These beliefs appear to have created and fed a false sense of security that the Third World nuclear states would be controllable. While attacks on the United States and other highly industrialized states are yet to become a reality, significantly improved delivery capabilities are spreading to the lesser developed states. While these delivery systems will not directly confront the industrialized states in the near future, they may destabilize the balance of power in other sections of the world.

Argentina is developing the medium-range Alacron (Condor II) ballistic missiles, with a range of 500 miles. While the Alacron, as such, will not be an intercontinental missile, yet it is a product of the Third World that will be exportable. Argentina has been linked with unconfirmed Alacron exports to Iran and Egypt, part of the network of Third World arms sales already threatening the control held by traditional arms control suppliers. Brazil is also developing two medium-range ballistic missiles, possibly for export to Argentina but more likely to other Third World states.[65] The Brazilian-Argentine attempts at subregional integration, illustrated by President Sarney's visit to the enrichment facility at Picaniyeu in 1987, show the level of growing trust between the states.

For Argentine use, the intercontinental ballistic missile under development by India might offer interesting possibilities. While the Indian ICBM has had several failures and is far from ready for deployment or sale, its existence serves notice of the intentions to arm themselves on the part of states outside of the traditional ranks of military superpowers. The new arms producers will likely contemplate the role that China and Brazil have had in changing the balances of Third World conflicts, and it is ludicrous to believe that emerging arms producers would not emulate the sales behavior of the more established sellers. Argentina could well become a recipient as well as a producer of the new weapons—a infinite danger even with nuclear weapons unsophisticated by U.S.-USSR standards.

The most pressing issue for Argentina is not the question of nuclear weapons production. Regimes back to the 1950s have consistently argued that the sole intent of the CNEA, and the government in general, is the self-sufficiency of the Argentine nation. As a result of the 1982 conflict in the southern cone, regardless of who started it, calls for Argentina to defend itself against Britain have grown to include some discussion of nuclear weapons. As long as Argentina remains outside the Treaty of Tlatelolco (1967) and the Non-Proliferation Treaty (1968), it appears possible, if not likely, that Argentina will exercise the nuclear option. A more credible scenario, however, has Argentina sticking to a position of ambiguity—not choosing to pursue weapons openly, yet retaining the national option for some point in the future. It will certainly maintain the option of developing nuclear weapons to defend the nation if necessary, even though no enemy clearly exists.

Argentina is more likely to begin selling international technology to the Third World, with perhaps deeper involvement, given its development of certain technologies not necessary for the current nuclear reactors. The 1983 acknowledgement of the Picaniyeu enrichment facility, developed in a clandestine manner, is one example. With the current heavy water reactors, there is no need for enriched uranium, and any plans to reorient its nuclear generating grid toward lightly enriched uranium/natural water systems seem highly unlikely in view of Argentina's commitment to a heavy water system. Since Argentina has not signed the major instruments of nuclear nonproliferation, it is not committed to refrain from selling *indigenously* built technology to other nations. If traditional suppliers begin adhering to the London Suppliers' Club (or some other) standard intended to curb sales of nuclear technology to states not part of the nonproliferation regime, Argentina might fill in the gaps. Its declared policy is that Argentina imposes its own bilateral nonproliferation standards on any nuclear sales. The nation's position has been one of sharing technology with others from which it can benefit, such as Brazil or the People's Republic of China. Argentina might reach to broader markets, however, where it would realistically have little influence on the use of nuclear technology. The Iran-Iraq war, with its arms supplies from a vast number of heretofore unlikely sources, exemplifies this possibility. As long as a civilian regime retains control over the Argentine state, the imposition of Argentine safeguards will continue to promote the peaceful uses of the nuclear energy.

Argentina has made the first major sales of nuclear technology to other

Latin American states. Through the Organization for Latin American Energy Development, as well as under bilateral agreements, Argentina has assumed a leadership role in the region. The Peruvian agreements of the 1970s to buy small research reactors were important for their symbolism as well as the reality that Argentina was able to provide indigenously produced technology—not nuclear materials bought from industrialized states and reexported. Discussions about selling small research nuclear reactors have taken place between Argentina and virtually every other Latin American state. These examples of regional cooperation, however, in nuclear as well as conventional energy ventures, only foreshadow the possibilities in the years to come. Regional ties appear to be strengthening in response to outside influences, such as those of the international banking community.

Argentina has achieved relative self-sufficiency in energy through petroleum, hydropower, and nuclear power. The question is whether it will have to sacrifice long-term self-sufficiency to meet short-term economic demands. The problems in getting the state-owned petroleum industry onto a sound economic footing show the difficulty facing all sectors of the energy field. Perhaps more than any other threshold nuclear weapons state, Argentina has the resources to make its own choices. Many of the problems that it faces pit domestic control over energy resources against the desire to exploit those resources more economically. Unlike several other threshold states, Argentina has sizable natural resources, which it is exploiting at a quickening pace in a diversified manner. It has no genuine security threat like those facing other states in this book, yet Argentina has chosen to pursue policies that perpetuate doubts regarding its status in the international system, specifically in nuclear development.

Of the threshold states, Argentina has the least reason to develop nuclear power, aside from maintaining its role as a potential weapons state. The problem with such a position is that the internal political turmoil of the past quarter of a century has prevented Argentina from creating a solid, sustained energy program. The history of the upheavals in its energy program, as well as those across society, has created doubts that a sophisticated civilian nuclear power program could escape being dragged into domestic conflicts in the future. Plutonium generated by the nuclear reactors might be used for nuclear weapons production whether Argentina remains a democratic state or reverts to some sort of authoritarian regime. In any severe domestic upheaval, even a stockpile of crude nuclear weapons could have a major impact on the internal and external affairs of the nation. While no absolute figures on plutonium production are available, indications are that Argentina has—or can have almost immediately—

sufficient fissile material to create weapons if the political will to do so is there.

It is more credible, however, that any regime in Buenos Aires will continue working to establish Argentina as a major supplier in the second tier of nuclear supplier states. This state of affairs makes the nation more visible as a player while at the same time offering some slight remedy for economic needs. Of real value in the short term of the ability to supply nuclear technology will be that it carries political weight through the perception of economic and technical sophistication.

Given the overriding concern about debt service, Argentina's national resources are being diverted from what the Menem government and its predecessors declared as their intent over the past thirty-five years. It is possible, however, for a Buenos Aires government—particularly if it returned to either the more traditional Perónist position or that of an insecure National Security State—to determine that national priorities are more important than those of the international banking community. In that case, Argentina might pursue its energy programs more directly, without considering the economic budgetary ramifications, and a decision to take a more concerted approach to developing nuclear weapons—or at least maintaining the nuclear ambiguity—would be in order. The questions for Argentina unlike those for other threshold states, lie largely within the country itself and are thus not easily resolved.

Regardless of whether it sells technology, develops its own weapons, or abandons nuclear energy altogether, the Argentine position will put national interest above anything else. While it will work to create self-sufficiency, creative methods may be invoked, depending on who is in control of the regime.

NOTES

I want to extend thanks to David Schier and William Stant of Loyola University of Chicago, David Nordstrom of DePaul University, and Etel Solingen of UCLA, who all participated in the development of this chapter.

1. Juan Domingo Perón, *Perón Expounds His Doctrine* (Buenos Aires: n.p., 1948), p. 243.

2. See the episode described in James Spiegelman and Peter Pringle, *The Nuclear Barons* (New York: Avon Books, 1981), or Joseph Page, *Perón: A Biography* (New York: Random House, 1983).

3. Fariboz Ghadar, *The Petroleum Industry in Oil-Importing Developing Countries* (Lexington, Mass.: Lexington Books, 1983), pp. 105-7.

4. A detailed study of the Argentine petroleum industry appears in Carl E. Solberg, *Oil and Nationalism in Argentina: A History* (Stanford: Stanford Univ. Press,

1979). For the 1920s and 1930s, see George Philip, *Oil and Politics in Latin America* (Cambridge: Cambridge University Press, 1982), chap. 8.

5. Ghadar, *Petroleum in Developing Countries,* p. 109.

6. Nazli Choucri, *Energy and Development in Latin America* (Lexington, Mass.: Lexington Books, 1982), p. 153.

7. Ghadar, *Petroleum in Developing Countries,* p. 111.

8. *New York Times,* Sept. 27, 1958, p. 26.

9. Ibid., April 21, 1957, p. 26.

10. *Review of the River Plate,* March 11, 1958, pp. 17-21.

11. *New York Times,* August 2, 1957, p. 24; July 25, 1958, p. 5; August 15, 1958, p. 35; Dec. 5, 1958, p. 50; Jan. 6, 1959, p. 42.

12. Ibid., May 12, 1961, p. 8; Sept. 27, 1961, p. 27; 27 December, 1961, p. 4.

13. *Latin American Weekly Report,* April 5, 1985, p. 10.

14. For discussion of the Soviet role in Latin America, see Cole Blasier, *The Giant's Rival* (Pittsburgh: Univ. of Pittsburgh Press, 1985). See also Aldo Vacs, *Discreet Partners* (Pittsburgh: Univ. of Pittsburgh Press, 1984), which discusses the growth in Soviet interaction with Argentina after 1970.

15. James Everett Katz and Onkar Marwah, eds., *Nuclear Power in Developing Countries* (Lexington, Mass.: Heath & Co., 1982), pp. 84-85.

16. *New Times,* Dec. 11, 1963, pp. 5-7, and *New York Times,* Nov. 17, 1963, p. 1.

17. Ione S. Wright and Lisa M. Nekhom, *Historical Dictionary of Argentina* (Metuchen, N.J.: Scarecrow Press, 1978), p. 1042, and *New York Times,* Feb. 10, 1965, p. 53.

18. *New York Times,* August 15, 1965, p. 28, and Oct. 30, 1965, p. 54.

19. Ibid., March 29, 1967, p. 61; Feb. 23, 1968, p. 49; and Dec. 19, 1968, p. 41.

20. Ibid., June 11, 1967, p. 12.

21. Ibid., June 1, 1972, p. 61.

22. Ibid., August 28, 1974, p. 9.

23. Ghadar, *Petroleum in Developing Countries,* p. 123.

24. "Producción de otros combustibles," *Anuario Estadístico de la República Argentina, 1981-82* (Buenos Aires: Instituto Nacional de Estadístico y Censos, 1982), p. 531.

25. Daniel Poneman, *Nuclear Power in the Developing World* (London: George Allen & Unwin, 1982), p. 70.

26. Perón's relations with the United States indicated that nations did not adhere to "rules of international behavior" but "rules of manipulation." The history of ups and downs in the relations of Argentina and the United States during Perón's tenure has been told in several places, as in Joseph Page, *Perón: A Biography* (New York: Random House, 1983).

27. Poneman, *Developing World,* p. 72.

28. Katz and Marwah, *Nuclear Power in Developing Countries,* p. 86.

29. Ibid., p. 88.

30. Poneman, *Developing World,* p. 74.

31. *Guardian,* April 17, 1982, and *Listener,* April 22, 1982, probed these questions with an air of heightened concern.

32. Carlos Castro Madero, "El Plan Nuclear Argentino y su Proyección Futura," *Futurable* (1981), pp. 25-40.

33. Poneman, *Developing World,* p. 74.

34. Ibid., p. 75-78, and *International Herald Tribune,* August 8 and 9, 1981.

35. Katz and Marwah, *Nuclear Power in Developing Countries,* p. 89.

36. Ibid., p. 89.

37. Poneman, *Developing World,* p. 76.

38. Ibid., p. 77.

39. See Etel Solingen's chapter on Brazil in this book.

40. Joseph Pilat and Warren Donnelly, "An Analysis of Official Argentine Statements on the Purpose and Direction of Argentina's Nuclear Program," U.S. Congressional Research Service, May 10, 1982, p. CRS-6.

41. Castro Madero, "El Plan Nuclear Argentino," p. 56.

42. *Economist*, January 26, 1980, p. 26.

43. *Financial Times*, April 6, 1979.

44. *New York Times*, March 27, 1980.

45. Poneman, *Developing World*, p. 80.

46. Ibid., p. 80.

47. *Financial Times*, April 6, 1979.

48. See William Walker and Mans Lonnroth, *Nuclear Power Struggles* (London: George Allen & Unwin, 1983). The London Supplier Club was founded in 1975-76, largely at the insistence of Henry Kissinger and Gerald Ford, sparked by the Brazilian sale and the Indian detonation. The purpose of the group was to develop a set of guidelines that the major supplier states of nuclear materials and technology would abide by in selling to the Third World. Essentially, the group was determining the lowest common denominator to keep the threshold states contented but provide no mechanism for building a weapon. The original members were the industrialized states.

The LSC was not well received in the threshold states. States such as India and Argentina claimed this was yet another attempt to discriminate against Third World countries, treating them as a dumping ground for products useless at home and denying them full rights to whatever was being developed in the state of technology. The LSC was labeled a cartel, on the grounds that the intent was merely to raise prices, not to allay any real fear that these states would build military devices. Fueling this criticism, the LSC withheld its "trigger list" of materials and technology not to be sold to the threshold states on grounds that the sensitive technologies and materials might be used by threshold states for weapons development. The "trigger list" was published two years after it was agreed upon.

As became clear with Atucha II, and otheres, the LSC has not upsheld its guidelines over the years. Governments have changed since the Ford-Carter nonproliferation push of the 1970s, but not all have agreed to sell less.

49. Pilat and Donnelly, "Statements," p. CRS-10.

50. The *guerra sucia*, or dirty war, was conducted between late March 1976 and December 1983, when Raúl Alfonsín assumed power. While outside the scope of this chapter, the National Security State during these years allowed considerable government control over the activities of citizens, while limiting control over those of foreign economic interests, helping to create the $50 billion debt left to the democratic successor regime.

51. *Southern Cone Report*, Dec. 24, 1987, p. 8.

52. Ibid., Oct. 15, 1987, p. 8.

53. *Latin American Weekly Report*, March 3, 1988, p. 12.

54. *Times* (London), July 24, 1982.

55. *Latin American Weekly Report*, Oct. 16, 1986, p. 11, and August 6, 1987, p. 2.

56. Ibid., Jan. 6, 1984.

57. Ibid., April 5, 1985, p. 10.

58. Ibid., August 23, 1985, p. 2.

59. Ibid., Sept. 25, 1986, p. 8.

60. Ibid. June 15, 1984, pp. 6-7.

61. *Southern Cone Report*, Feb. 2, 1989, p. 8.

62. *Latin American Weekly Report*, Oct. 16, 1986, p. 11.
63. Ibid., May 26, 1988, pp. 4-5.
64. *New York Times*, July 22, 1987.
65. Ibid., July 3, 1988, p. E3.

7

Brazil

ETEL SOLINGEN

Brazil's involvement in the international economic system has increased remarkably since 1964.[1] The extent to which growing interdependence has diminished or strengthened the country's ability to adjust to a broad range of external challenges may be gauged by various yardsticks. Among these, its experience in the energy sector provides a critical measure of state capacity to adapt to changes in the international political and economic environments. Why were certain instruments of domestic and foreign policy selected and not others?

The first task is to define the nature of the threat to political coalitions and state structures posed by the 1973-74 and 1979 oil shocks. Brazil's response to these challenges reflected the country's political and institutional configurations. Adjustment strategies were inextricably linked to the macropolitical objectives of the military-technocratic regime installed in 1964. The goals included rapid economic growth through the accelerated integration of Brazil into the international economic system. The expansion of state entrepreneurship, high levels of external indebtedness, and import substitution were core instruments in this strategy, which at times undermined the attempt to strengthen a national private industrial base. Thus, the strategy was not exempt from contradictions that helped to erode the political basis of the ruling coalition. In particular, domestic private entrepreneurs denounced the expansion of the state sector. The regime's domestic weakness influenced the nature of immediate foreign policy responses. These took the form of an accommodating web of economic and political ties with oil suppliers and an emphasis on diversifying and deepening Brazil's relations with the industrialized world. Structural changes in the international system—including Brazil's emergence as a rapidly industrializing country—reinforced a policy of moderation. The characteristics of Brazil's nuclear program further reflected the major parameters in domestic and foreign adjustment strategies. Some of the ambiguities in Brazil's strategies were shared by other newly industrializing countries that placed energy resource scarcities at the core of their regional and global policy.

Table 7.1. Energy and the economy in Brazil (1970-1982)

	Real GNP growth	Price increases for petroleum and derivatives	General price index	Trade balance (billions)	Foreign debt (billions)
1970	8.8%	17.7%	19.8	U.S.$ 0.2	U.S.$ 5.3
1971	13.3	26.5	18.7	− 0.3	6.6
1972	11.7	23.8	16.8	− 0.2	10.2
1973	14.0	14.7	16.2	0.007	12.6
1974	9.5	65.5	33.8	− 4.7	17.4
1975	5.6	52.4	30.1	− 3.5	22.0
1976	9.7	57.6	48.2	− 2.3	29.0
1977	5.4	39.3	38.6	0.1	32.0
1978	4.8	30.2	40.5	− 1.0	43.5
1979	6.8	67.8	76.8	− 2.7	49.9
1980	7.9	159.4	110.2	− 2.8	54.0
1981	− 1.9	120.9	95.2	1.2	61.8
1982	1.3	71.5	99.7	0.8	69.7

Sources: Inter-American Development Bank, Economic and Social Progress in Latin America, 1988; W. Baer, The Brazilian Economy (New York: Praeger, 1983).

THE NATURE OF THE THREAT

In the 1970s, no single event had a greater impact on Brazil's domestic and foreign policies than the energy crisis—perhaps one of the most important sources of foreign indebtedness, which by the early 1980s had helped turn the country into the largest debtor in the Third World. The "boom" of 1969-73, with its 11 percent annual growth rates, was based on the expansion of manufacturing, which was export-oriented and dependent on cheap energy. Table 7.1 highlights some of the major parameters of economic performance between 1970 and 1982 and the nature of oil price increases during that period.

Energy consumption patterns generally favored petroleum—85 percent of it imported—which accounted for 41 percent of Brazil's total energy requirements in 1972. In the wake of the 1973 oil price increase (1973-83), Brazil had to set aside over 30 percent of its export earnings to pay for foreign oil, thus turning an export-oriented strategy of economic growth into an export drive to meet import needs.[2] The origins of the country's current economic crisis can, to some extent, be traced to the pursuit of the fundamental objectives of ensuring energy supplies and softening the political and economic effects of dependence on foreign sources. The first wave of price increases helped transform a bare trade surplus of $7 million in 1973 into a $4.69 billion deficit in 1974 (tables 7.1

Table 7.2.　Cost of Brazil's oil imports 1971-1987 (selected years)

	Average cost per barrel (FOB)	Total cost of oil imports (billions)	Cost of all imports (billions)	Oil as percentage of imports
1971	U.S.$ 1.88	U.S.$ 0.4	U.S.$ 3.2	11.6%
1972	1.99	0.5	4.2	11.1
1973	2.79	0.8	6.2	12.4
1974	11.11	3.0	12.6	22.9
1975	10.49	3.1	12.2	25.2
1976	11.50	3.8	12.4	30.9
1977	12.30	4.1	12.0	33.8
1978	12.44	4.5	13.7	32.8
1979	17.11	6.8	18.1	37.3
1980	30.60	10.2	23.0	45.0
1981	34.37	11.3	22.1	51.0
1982	33.00	10.5	19.4	53.0
1983	28.00	8.6	15.4	56.8
1986	13.00	2.7	13.0	30.0
1987	17.1	3.8	15.0	25.6

Sources: George Philip, *Oil and Politics in Latin America* (Cambridge: Cambridge University Press, 1982) p. 389; *Brazilian Monthly Economic Indicators*, December 1983; *Economic and Social Progress in Latin America*; W. Baer, *The Brazilian Economy* (1988).

and 7.2). Higher oil prices, however, accounted for only $2 billion of the total deficit.

On the one hand, Brazil's energy priorities were linked to the most fundamental objectives of the military-technocratic coalition that took power in 1964—namely, rapid economic growth and national security. President Castello Branco defined national security as "the preservation of development and internal political stability," and energy policies lay at the core of developmental priorities.[3] On the other hand, the inflationary pressures of the energy crisis compelled the deceleration of growth targets in 1976 and 1979. The social and political corollaries of these strategies were costly for an authoritarian regime that sought domestic legitimacy through sustained economic growth. The threat, in other words, was not merely economic but a challenge to the stability and legitimacy of a political model. Preserving large-scale productive capacity in energy-dependent intermediate goods sectors such as cement, petroleum refining, petrochemicals, steel, and aluminum, where the state had a commanding position, was a central objective. Public enterprises subsidized inputs to other sectors, including private enterprise, and provided quasi-monopsonistic markets for capital goods, engineering, and other products and services.

In short, the energy shocks had the potential of undermining the political and economic basis of support for the "model." The salience of energy policies was reflected in personal and institutional adjustments as Ernesto Geisel, a former head of the state oil firm Petrobrás (1969-73), assumed the presidency of Brazil in 1974. Energy policies in general, and the nuclear program in particular, were subject to widespread criticism and polarized the scientific community, the technocrats, and the generals themselves.

CHANGES IN DOMESTIC STRUCTURES

The selection of crisis adjustment mechanisms was influenced more by macropolitical objectives than by the opportunities offered by Brazil's natural energy endowments. Import substitution, export expansion, and external indebtedness were the dominant strategies of adjustment, allowing sustained economic growth between 1974 and 1978, albeit at lower and more variable levels than during the preceding five-year period.[4] Import substitution in capital goods, petrochemicals and derivatives, steel, metal products, and energy opened up unique opportunities for domestic industry, but it was largely sustained by external borrowing. International financial markets made massive borrowing attractive as an initial response aimed at protecting the growth structure and the continuity of the political alliance in place since the 1960s. To avoid curtailing domestic consumption, domestic oil prices were not raised to world levels. Immediate income declines were not politically feasible, particularly in the delegitimizing environment reflected in the 1974 elections, which doubled the strength of the opposition in the Chamber of Deputies. This outcome was of particular concern to President Geisel who had recognized the need for political liberalization (distensão).

External challenges forced the restructuring of domestic priorities and enhanced the position of state firms in the productive and financial areas. The twin aims of securing supplies and promoting energy independence were pursued through policies of substitution for oil through the development of alternative energy technologies, such as hydroelectrical, nuclear, alcohol, biomass, and coal. The energy crisis thus led to a broader reformulation of industrial policy affecting the capital goods, engineering, construction, and other industries. The following brief survey of these changes in the energy sector will place particular emphasis on patterns of accommodation between domestic and foreign resources. It will also examine the extent to which state behavior was market-displacing or market-conforming.[5]

The expansion of the state as an economic agent throughout the energy sector, much as in other areas of the economy (mining, steel, petrochemicals) was expressed in the prominent role played by state energy companies, particularly Nuclebrás, Petrobrás, and Eletrobrás. Electric power generation shifted from private to public hands in a single decade: State enterprises, which controlled less than 36 percent of power-generating capacity in the 1960s, controlled over 80 percent in the late 1970s, and close to 100 percent by the early 1980s.[6]

Petrobrás, Brazil's largest economic concern, was set up in 1954 to refine imported oil and generate financial resources to achieve self-sufficiency. It controlled petroleum exploration and imports and expanded into distribution and related new fields—often competing with private firms—through subsidiaries like Petroquisa (petrochemicals), Petromisa (minerals), and Braspetro (foreign oil prospecting and technical assistance).[7] Rather than turning to private-sector firms for transportation and construction, Petrobrás created its own subsidiaries. For over 85 percent of its total capital goods requirements, the firm turned to domestic suppliers.[8] Despite a long tradition of nationalist objectives expressed in the creation and evolution of Petrobrás, in an unprecedented reaction to the oil crisis Brazil granted exploration rights to foreign companies in 1975. Under "risk contracts," which ultimately bore little fruit, exploration was allowed in exchange for a share of the oil discovered, thereby undermining the historical monopoly of Petrobrás.

Eletrobrás is responsible for hydroelectric power, which accounts for 95 percent of electricity generation. Inducements for developing this potential include the availability of rivers as clean, nonpolluting, renewable resources, as well as the possibility of significant reliance on domestic capital goods and engineering firms. Although it controls the biggest hydroelectric reservoir in the world (70,000 Mw), Brazil has utilized only about 15 percent of its generating capacity. The hydroelectrical component in the energy balance has increased considerably, from 16.8 percent in 1969 to 28.3 percent in 1979, and 38 percent in 1985.[9] Emphasis on the Amazon basin grew after the debacle over the nuclear program in the 1970s, and in 1984 the world's largest hydroelectric project, Itaipu Binational, was inaugurated. The environmental effects of some of the projects have involved Brazil in major controversies with environmental groups in both the North and South.[10]

Large-scale projects using advance technology and imported machinery, with heavy funding from the World Bank and Inter-American Development Bank, were the norm in the 1960s and 1970s.[11] Thirty-three major plants, each with a capacity of nearly 1,000 Mw, were being built or enlarged in the early 1980s.[12] Reflecting the emphasis on domestic supply of capital goods, 80 percent of inputs to Eletrobrás came from local sources. The National Bank for Economic and Social Development (BNDES) provided effective support for projects achieving a domestic share of more than 85 percent. On the other hand, suppliers' credits and relaxed import policies in the areas of turbine and hydro generators contributed to the high levels of idle capacity at national firms in this sector, exacerbating their discontent.

Since Brazil depended on oil less for electricity generation than for transportation and for consumption in the commercial and residential sectors, alcohol proved an attractive substitute.[13] Brazil pioneered in the development of alcohol fuels—methanol, obtained from coal or biomass gasification, and ethanol, from fermentation of sugar cane. By the mid-1980s, 7 million cars in Brazil, and 80 percent of all new Brazilian cars, were powered by a gas mixture containing 20 percent alcohol. In addition, over 1 million cars used hydrated alcohol as their exclusive fuel. State efforts in the alcohol sector were directed at strengthening private Brazilian entrepreneurs in the areas of sugar cane production, distilling, retail fuel distribution, capital goods supplies, and automobile manufacturing.

Advocates of alcohol fuels pointed to the savings in foreign exchange, the renewable quality, low environmental and transportation costs, generation of employment and income, reduction of individual and regional income disparities, and stimulation of the domestic industry through the production of new refinery equipment.[14] However, the eroding price advantage of ethanol-fueled cars and the growing social cost of the alcohol program, are giving way to renewed demand for oil-fueled automobiles. The alcohol program became a test case for the debate over increased state expansion, with private domestic and multinational firms challenging Petrobrás's aim to retain control over the liquid fuels sector.[15] In the long run, alcohol did little to alleviate demand for oil imports.

Coal accounts for only about 9 percent of Brazil's total energy needs; 40 percent of it comes from domestic sources. Estimates suggest that coal reserves will last well into the next century—even with a projected consumption level about forty times greater than coal use in the mid-1980s, and about 2.5 times higher than the current total energy consumption in Brazil.[16] The state firms Siderbrás (iron and steel), Petrobrás, and Caeeb market gasified coal at subsidized prices. In an attempt to substitute coal for fuel oil in industry, state intervention in this area was limited to providing transitional subsidies to coal producer and consumer sectors (notably cement firms), and to improving the transportation infrastructure between the two. Charcoal and firewood represented 35 percent of the primary fuel consumed in 1972, decreasing to 20 percent in 1980, when charcoal contributed only 2.5 percent. Exploitation of these resources bears some responsibility for the devastation of Brazil's natural forests.

Being a tropical country of vast proportions, Brazil has a great potential for solar energy. Similarly, the country is endowed with some of the largest oil shale reserves in the world, although research on extracting the processing technologies has not been given high priority. Other biomass alternatives—alcohol derived from eucalyptus, sugar sorghum, manioc— have been advocated because of their renewability and small capital investment requirements, their labor-creating potential in rural areas, and the absence of waste disposal problems. As very little oil was being used for electricity generation, nuclear power offered no real substitute for oil in the energy crisis of the 1970s. Yet an agreement was signed with West Germany in 1975 providing for the transfer of eight nuclear plants and the complete fuel cycle, stressing self-sufficiency and technological advancement. (A more detailed analysis of the nuclear program follows.)

The oil price shock in 1979 forced a new series of adjustment strategies. The crisis was compounded by rising interest rates that aggravated the debt service outflow dominating the current account deficit, and by the world recession of 1980-82.[17] A concomitant erosion of the regime's political legitimacy accelerated when powerful industrial sectors assumed the leadership of a campaign against centralization and the expansion of state activities. The steps taken between 1974 and 1978 had not been successful in preventing an erosion of the 1968-73 model; the average growth rate of the pre-1973 period declined toward the late 1970s, leading to the three-year recession of 1981-83. President Joao Batista Figueiredo warned in 1979 that the oil crisis could compromise the country's stable development and international credibility, and he called for increased production of domestic oil, coal, and ethyl alcohol from sugarcane.[18] Hydropower was to have priority over nuclear generation, previously an untouchable, privileged item. While energy policies before 1979 were incoherent and unintegrated, the second oil shock led to a more comprehensive approach to energy alternatives.

The statistics suggest a marked qualitative change in the structure of oil dependence in Brazil in the last ten years. The volume of imports dropped from 950,000 barrels per day (bpd) in 1979 to 620,000 bpd in the first quarter of 1983. Domestic oil production grew from 340,000 bpd in 1983 to 609,000 in 1988. The cost of imported oil fell from $8 billion in 1982 to less than $4 billion in 1987 (see table 7.2). Dependence on foreign sources dropped from 84.4 percent in 1979 to 73.7 percent in 1982, then to an estimated 45 percent in 1988.[19] Petroleum's share of total energy requirements fell from 44 percent in 1975 to 24 percent in 1985. The relative success in oil conservation can be attributed to the substitution of alcohol fuels and coal. Yet oil imports continue to account for about 30 percent of Brazil's total energy requirements in the late 1980s, up from 25 percent a decade earlier. Despite diversification, increased domestic production, a drop in international oil prices, and growth in Brazilian manufactured exports to the Middle East, a negative balance of trade with that region persists into the late 1980s.[20]

In early 1982, President Figueiredo approved Plan 2000, which called for a slowdown in development of all nuclear facilities. The Iguape reactors (third and fourth of the eight considered in the agreement with West Germany's Kraftwerk Union) were indefinitely postponed.[21] The alcohol program, instead, received significant governmental incentives. The projected 11.8 percent annual GDP growth rate was revised, since the economy, which had expanded at an annual average of 6.5 percent between 1975 and 1979, declined dramatically in 1981-82, and electricity consumption increased at only half the rate predicted.[22] More recently, the new Brazilian constitution approved in 1988 nationalized exploration for, and extraction of, oil and minerals—sectors where foreign firms had previously been allowed to operate.[23] It also required oil companies to assume the financial risk of exploration, previously subsidized by the state.

The direct effects of energy planning on the foreign debt crisis are expressed in the considerable portion of the public debt accounted for by state firms in the energy sector. Petrobrás, Eletrobrás, Siderbrás, and Cia Vale do Rio Doce had relied on private foreign banks for 17 percent of their investment in the early 1970s, with internal resources accounting for 30 to 50 percent. The level of self-financing dropped to about 25 percent in 1980. Electrical utilities depended on foreign resources for about 30 percent of their total borrowing.[24] Eletrobrás alone was responsible for $12 billion of the total debt in 1985. Its subsidiary Furnas had a $2.6 billion foreign debt in 1983, and that of Nuclebrás approached $4 billion in 1988.[25] The public sector as a whole accounted for close to 80 percent of the $120 billion foreign debt in 1988. IMF conditionality schemes and other public and private creditors have applied pressure to reduce state enterprise deficits.

Two tendencies stand out in Brazil's domestic response to the oil crises of 1973 and 1979. First, displacement of private sector firms was more characteristic of state behavior than were market-conforming efforts, such as those in the alcohol sector, petrochemicals, and others, geared to encourage private sector participation. Subsidiaries of Petrobrás, Nuclebrás, and other state firms established as joint ventures with foreign partners became the most common institutional expression of this effort. Heavy state ownership in energy markets, usually justified as a function of the nature of the investment and private sector reluctance, is not unique to industrializing countries. Ownership is sought, in part, because it provides greater national control over volatile international markets, thus reducing external vulnerabilities. The state's ability to maintain control is strengthened by its comparative advantage over private interests in the conduct of foreign policy, by the high degree of standardization of energy markets, and by these markets' role as upstream suppliers of other industries.[26]

Second, most of the investments of this period were possible because of the thriving financial markets of the 1970s, particularly in Eurodollars. These markets were the pillars of Brazil's initial refusal—through price controls and other cushions—to allow rising oil prices to impinge on the continued growth of basic industries and infrastructure. However, as Albert Fishlow's analysis of Brazil's management of the oil crisis suggests, foreign-financed public investments were more a sign of state weakness than of its strength, and continuous reliance on external indebtedness reinforced that state of affairs. Against the contradictions of the domestic background, we turn now to the international expression of energy adjustments and to the instruments of foreign policy adopted to manage a new international environment.

ADJUSTMENT IN FOREIGN POLICY INSTRUMENTS

Economic growth, or a rising share of the world's GNP, is assumed to increase state power and upward mobility in the international arena.[27] Brazil's share of world income increased from 1.16 percent in 1967 to 1.77 percent in 1976 and 2.7 in 1988.[28] Moderate as they may seem, these changes have placed Brazil on par with Australia, Canada, and some smaller European countries with respect to contribution to the world's GNP. Yet, whether because of internal political and economic weaknesses or external vulnerabilities such as dependence on foreign energy, capital, and technology, the country has exercised only moderate international influence relative to its growing capabilities. This disparity is particularly evident in its accommodating reactions to the OPEC onslaught.

Brazil was not as active as it could have been in encouraging economically powerful but oil-poor industrializing states to coalesce in a common response to OPEC.[29] It did not seek to join or promote a consumers' cartel, stayed away from military solutions advocated by a few at the time, and opted for extensive borrowing to finance oil imports. It mildly encouraged multilateralism while pursuing aggressive bilateral arrangements. It resisted attempts to upgrade its international credentials in ways that would impose burdens in the name of responsibilities. Brazil's foreign policy was conciliatory, determined more by considerations of economic growth than by the desire to exert political influence. As Selcher suggests, the "business of Brazilian foreign policy *is* business."[30] Brazil's ruling coalition regarded the achievement of commercial competitiveness, the attraction of foreign investments and technology, and the securing of energy supplies as primary strategies, requiring a cautious and restrained performance. The guiding principles of "responsible pragmatism" and "no automatic alignments" were based on the conception that "material efficiency rather than formal coherence is the standard of policy evaluation."[31]

The notion of a fundamental interdependence between Brazil's foreign policy and energy considerations is the underlying premise of the brief assessment of global and regional diplomacy that follows. The view that foreign policy, if wisely formulated and skillfully applied, could produce economic benefits guided the military-technocratic regime from its inception in 1964. Diplomatic action aimed at the expansion of foreign markets and the attraction of foreign investments became popularly known as "the diplomacy of prosperity." The 1973 energy crisis reinforced this preoccupation with economic issues and brought about a reassessment of Brazil's position vis-à-vis OPEC countries, the rest of the Third World, the advanced industrialized countries, and its own neighbors. The shift to economic diplomacy was accompanied by institutional changes. Many of the tasks previously performed by the Foreign Ministry (Itamaraty) were transferred to the Ministry of Mines and Energy, the Ministry of Industry and Commerce, the Ministry of Finance, and the state companies Petrobrás, Nuclebrás, and Eletrobrás.

A major foreign economic policy instrument to offset the impact of oil imports on the balance of trade was the promotion of exports to oil suppliers. Interbrás was created to promote and sell Brazilian manufactured goods and commodities and even barter them for energy resources. By 1978, Brazil had multiplied the value of its 1972 exports to the Middle East and Africa by a factor of ten. It was among the world's top ten manufacturers of weapons in the early 1980s, with over fifty countries providing a market for its armored vehicles, the Cascavel, Urutu, and Jararaca. The buyers include Iraq, Iran, and Libya, which had extended opportunities to test them on the battlefield, and more recently Saudi Arabia. Other military exports include the Astros II rocket and the Tucano trainer. Arms and equipment sales of over $1 billion in 1982 represented an increased share of all Brazilian exports.[32] Brazil's attractiveness as a weapons supplier stems from the simplicity and effectiveness of its technology and the lack of ideological strings attached to the sales.

In addition to raw materials and manufactured goods, Brazil has pursued an aggressive policy of exporting technology and services, particularly in oil exploration and construction. By 1976, the Petrobrás subsidiary Braspetro had negotiated agreements with and successfully drilled wells in Iraq, Iran, Algeria, Angola, Congo, and Colombia. It received authorization from the Nicaraguan government for oil exploration and research, and it signed a risk contract with South Yemen. The Petrobrás trading company Interbrás provided engineering, financing, and commercial services to oil suppliers particularly, mostly in government-to-government transactions.

Middle Eastern countries remain Brazil's main oil suppliers, accounting for about 75-80 percent of total imports in 1986, with Saudi Arabia and Iraq respectively 30 percent of the total furnishing. Iran, the United Arab Emirates, Qatar, Libya, and Kuwait provided another 18 percent.[33] The outbreak of the Iran-Iraq war brought about the loss of 400,000 barrels per day in 1980, signaling the importance of Venezuela and Mexico as alternate suppliers. In 1981, Brazil signed three energy agreements with the USSR for coal technology and the production of methanol alcohol from wood. Yet by the mid-1980s, the Soviet Union, Venezuela, Ecuador, and Mexico were supplying only a little over 4 percent of Brazil's oil imports. The price advantage of Middle East oil was the justification for the shift away from Venezuelan oil during the 1960s, and it continues to be at the root of dependence on Middle Eastern suppliers. Negative trade balances with the Middle East persist into the late 1980s.[34]

One of the political and diplomatic expressions of these economic relationships has been, since 1973, Brazil's movement away from a position of "equidistance" in the Arab-Israeli conflict. Accordingly, it supported the 1975 United Nations anti-Israeli vote equating Zionism with racism, while eighteen other Latin American countries either casted a negative vote or abstained.[35] Relations with Libya's Muammar Kaddafi provide a striking example of policy reversal, from negligible links before 1973 (when Kaddafi sponsored Brazilian antiregime exiles) to close military and commercial ties. The results of this strategy of cajoling oil suppliers were mixed, despite a modest increase in Brazilian exports to them, since no advantage in oil prices was gained, and the expected petrodollar investments by Arab countries did not materialize.

In Africa, Brazil was among the first countries to recognize Angola's Marxist (MPLA) government, headed by Agostinho Neto, in 1975. President Geisel hoped that Petrobrás would play a part in the exploitation of the Cabinda oil fields, a matter discussed with the MPLA before recognition.[36] Rather than stressing openly the strategic-economic potential of alliances with oil-producing countries such as Angola, Nigeria, and Gabon, or of securing a foothold in the little-tapped consumer markets of Africa, Brazil used the commonalities of the Portuguese colonial tradition as vehicles to strengthen its association with former Portuguese colonies in Africa. The pragmatic approach dictated by commercial considerations is again evident in the growth of bilateral trade with South Africa, from $7 million in 1972 to $150 million in 1980.

Adaptation to the constraints imposed by the regime's model of economic development is also reflected in Brazil's regional policies. Many concepts from its school of geopolitical thought have been incorporated into foreign policy principles.[37] Its essence is linking "security" and "development" as the military's core mission. Developmental goals, in turn, are related to the ability to ensure energy supplies. Issues of integration, the conquest of the Amazon heartland, and the South Atlantic Narrows are intertwined with priorities such as access to energy and natural resources.

In his seminal work *Projeção Continental do Brasil*, Colonel Mario Travassos argued in the 1930s that by projecting itself into the Amazon Basin and Bolivian heartland, Brazil could fulfill its "continental destiny." The prominent geopolitician and 1964-regime ideologue General Golbery do Couto e Silva, the first director of the National Intelligence Service and a former strategist at the Escola Superior da Guerra (Higher War College), conditioned Brazil's achievement of international influence on the following principles: regional and Third World cooperation, national integration, expansion into the interior, peaceful external projection, and participation in the defense of Western civilization. The Escola Superior da Guerra's doctrine of "development and security" has consolidated these theories and disseminated their essence through the socialization of the military-technocratic elite.[38] Its orientation has shifted away from cold war considerations and its influence has waned since the late 1960s.

These regional orientations provide a useful frame of reference for analyzing Brazil's relations in South America. Because national security is defined in terms of development, industrialization, and integration, strategic concerns vis-à-vis its neighbors reflect the pragmatism characteristic of Brazil's policies toward the rest of the world. In the regional context, Brazil has secured a paramount position through a series of agreements with the "buffer states" (Uruguay, Paraguay, and Bolivia), thereby creating a security perimeter and privileged area of economic activity. Central America and the Caribbean are of lesser concern to Brazil, even though countries like Guyana and Nicaragua have sought its help in developing their own versions of energy models.

Despite the hegemonic, expansionist labels often applied to its regional policies, Brazil has relied most often on negotiation, persuasion, and prestige, downplaying confrontation.[39] Not surprisingly, since 1973 the initiatives toward neighboring countries have been launched in connection with energy resource needs. An agreement with Colombia addressed the development of its coal reserves, another with Caracas provided for joint marketing of Venezuelan oil, while a tripartite agreement with Colombia and Venezuela dealt with development of the Amazon region. Other arrangements enabled Petrobrás to buy natural gas, petroleum, and electricity from Bolivia and to conduct oil exploration in Ecuador and Paraguay. Colombia provides Brazil with coal and is interested in Brazil's alcohol and uranium mining technology. In 1978, Brazil and Venezuela signed a $2 billion agreement to construct a 9-million-kilowatt dam in Venezuela's Guyana region, the world's third largest dam. In 1983, they signed a cooperation agreement in the area of nuclear energy for peaceful uses, under which Venezuelan technicians began training in Brazil. In 1982, an agreement was signed with Guyana concerning the hydropower station of Wamakuru.

The Amazon project launched by President Emílio G. Médici in 1970 was justified on the basis of economic, infrastructural, and rural developmental potential, but was deeply rooted in its perceived military-strategic relevance. Andean Pact countries regarded this project as destabilizing the previous geopolitical balance in an attempt to "project" Brazil's influence toward its northern borders.[40] Brazilian diplomacy (and Itamaraty's professionalism), however, succeeded in downgrading these perceptions of threat through a new series of bilateral treaties of friendship, trade, and cooperation, as well as the multilateral Amazon Pact Treaty of 1978. Regional projection has been accompanied by the settlement of about 80,000 Brazilians in Paraguay, northeastern Bolivia, and northern Uruguay through informal migrations and land purchases.[41]

Brazil's economic expansion into the buffer states eclipsed Argentina's influence. Brazil has four times Argentina's population and three times its GNP, historically more stable political leadership, and a more rapidly expanding economy. The territorial competition between the two dates from the early colonial expansionism of the Spanish and Portuguese crowns, which continued after independence in the early nineteenth century.[42] Two of the most critical issues in their friction during the 1970s were the two countries' respective nuclear and hydroelectric energy projects. This priority lends substance to the proposition that issues of development and integration, to which energy sources are subservient, set the tone in the countries' regional and global security and cooperation policies.[43]

Until 1960, Argentina, Uruguay, and Brazil agreed on a system of mutual consultations regarding their respective hydroelectric planning. In the late 1960s, Brazil expressed its reluctance to submit to the approval of the other countries of the Plata Basin what it considered its sovereign and unilateral right to exploit its hydroelectric potential. Yet in October of

1978, partly in view of vigorous Argentine protests, Brazil, Argentina, and Paraguay signed a tripartite agreement on the conditions for the operation of the Itaipu hydroelectric plant, to ensure navigation and safe water levels on the lower Paraná River. The settlement of the dispute over the exploitation of the Paraná in late 1979 opened the way for the 1980 visit to Argentina by President Figueiredo, the first by a Brazilian president in 45 years, which resulted in a series of agreements on hydroelectric and nuclear power, scientific and technological development, and various other economic and cultural issues.[44] The possibilities of Argentine exports of natural gas to Brazil and of joint petroleum exploration in Argentina were also explored. A series of presidential-level meetings, particularly between Raul Alfonsin and Jose Sarney, and, more recently, between Sarney and Saul Menem, has led to significant steps toward economic integration since the mid-1980s.

The role of Brazil in the South Atlantic has been discussed in the context of its technical and military capabilities and its economic goals.[45] Historically a major trade route, the South Atlantic became the vital petroleum lifeline for Europe and the United States with the closure of the Suez Canal in 1967. Supertankers continued to use that route even after the reopening of the canal in 1975. Considering Brazil's dependence on Middle Eastern and Nigerian oil and the increasing importance of West Africa and Angola as export markets, it is hardly surprising that even Brazil's South Atlantic strategies are conditioned by economic needs in general and energy considerations in particular. These interests seem to outweigh the advantages of the proposed South Atlantic Treaty Organization in cooperation with Argentina and South Africa, a proposal that has never been given serious attention. Brazil cannot afford to ignore South African racial policies, not only because of its own ethnic composition but also because of its commercial interests in Black Africa and has, therefore, shied away from any such agreement.

In its relations with the United States, Brazil has sought new options to deal with problems of development. Under the impact of the energy crisis, the trend toward greater autonomy has accelerated, putting an abrupt end to the "special relationship" that had existed between the two countries, buttressed by the experience of a joint brigade in World War II. Nuclear energy and trade became two major areas of friction in the 1970s, intensifying Brazil's search for new trade and investment partners, particularly among Western European nations and in Japan.[46] The Carter administration attempted to prevent, and later to disrupt, the implementation of the 1975 nuclear agreements with West Germany, which provided for the transfer of sensitive enrichment and reprocessing technologies to Brazil. In 1977, Brazil abrogated the 1952 military assistance treaty with the

United States as proof of its reduced need for foreign military supplies and its dismay at U.S. complaints over Brazilian human rights abuses. This move was designed to preempt an almost certain cut in U.S. military aid to Brazil. Commercial ties with Moscow, Angola, Cuba, and Libya have been not only financially lucrative but also a useful signal of independence. U.S.-Brazilian relations, however, are still intense, with the U.S. being the largest single-country market for Brazilian exports. Under the Reagan administration, disagreements over Brazil's nuclear program subsided, but Brazil's supply of weapons in the Iran-Iraq war and to Libya was high on the bilateral agenda.[47] Debt, trade, services, and technology remain the leading areas of dispute at the end of the decade.

At the level of global interactions, energy vulnerabilities may have constrained Brazil's transition to a more assertive participation in world politics. Efforts at rationing, oil exploration, diversification, and export policies were accompanied by an attempt to placate its preferred providers. Policies with respect to North-South issues were characterized in the 1970s by a reluctance to assume leadership positions and to give automatic support to raw materials cartels or regulation of foreign investment. As Selcher suggests, Brazil avoided taking a political lead in most major international issues because such issues are polarizing and have the potential of alienating "the diverse and demanding constituencies on which it depends."[48]

While Brazil's extraregional policies have attempted mainly to temper destabilizing impacts, within the regional context there has been greater willingness to exercise influence. The competition for control of economically important territory is at times considered a major source of tension in South America. However, through quiet and consistent diplomacy Brazil has been able to minimize regional conflict and at the same time carry out its developmental designs along its borders. Even though it declared Latin America to be a priority region for its national diplomacy, in practice Brazil's more pressing international economic commitments seem to take precedence. Its regional policies aim more at enhancing its influence through cooperation than at displaying hegemonic designs by challenging or threatening the neighboring countries. Yet there is a tacit source of threat—its nuclear capabilities.

NEITHER 'PYGMY' NOR 'PARIAH': BRAZIL'S NUCLEAR POLICY

Although stimulated by the 1973 energy crisis, Brazil's plans to acquire atomic power can be traced to 1951, when the National Research Council (CNPq), under the direction of Admiral Alvaro Alberto, assumed control of the nuclear sector. Stressing autonomous development and limitation of mineral exports, President Getúlio Vargas in 1952 approved directives to the National Security Council calling for "specific compensations"—technical aid and delivery of equipment and materials—in return for sales of uranium and thorium to the United States. However, following the ascendancy of President João Café Filho in 1954 and his dismissal of Admiral Alberto, two nuclear agreements were signed transferring the monopoly over uranium research and extraction to the United States. In the same year, Washington prevented the transfer of ultracentrifuge enrichment equipment from Bonn to Brazil.

Presidents Jânio Quadros (1961) and João Goulart (1961-64) encouraged national research and control of resources and diversification of external sources of technology. Translating these political guidelines into technical options meant reliance on natural uranium, and French natural uranium reactors were viewed as a possible option. At this point, there was agreement between nationalist sectors and the scientific community as to the nature of the technical path to be pursued.[49] The military coup in 1964 marked the beginning of a new phase, favoring reliance on imported fuel and technology. By 1968, the decision was made to opt for a pressurized water reactor (PWR) of the light water type, later purchased from Westinghouse as a turnkey. The decision in favor of enriched uranium was coupled with a purge of physicists in research centers who favored a more autonomous program based on natural uranium, thorium, and domestic technology.

In 1975, Brazil embarked on an ambitious attempt to master the entire nuclear cycle through an agreement with the West German firm Kraftwerk Union (KWU). The comprehensive arrangement provided for the transfer of eight nuclear plants of 1,200 Mw each; mining and uranium processing activities, including enrichment; plutonium reprocessing; and a joint venture in heavy components fabrication (table 7.3).[50] The most prominent arguments favoring this course included fulfillment of energy needs in the post-2000 era, reduced dependence on foreign sources of fuel, and the presumed multiplier effects of a nuclear industry. The light water, enriched uranium path was justified through a mixture of economic and technological advantages, notably its lower cost and greater technical reliability when compared with alternative cycles. Mastery over the entire fuel cycle would prevent dependence on imported fuel supplies, of particular sensitivity because the United States had ceased its transfers of enriched uranium to Brazil in 1974.[51] Brazilian industries would take over an increasing share—up to 90 percent by 1990—of both power plant construction and components manufacture.

The agreement with KWU was criticized in Brazil on several grounds. The jet nozzle enrichment procedure was considered a great risk, as the

Table 7.3. Nuclebrás and subsidiaries

	Activity	Ownership
Nuclebrás	Fuel cycle Power plants Technological research	100% Brazilian government
Subsidiaries		
Nuclam	Uranium prospection	51% Nuclebrás 49% Urangesellschaft (FRG)
Nuclei	Isotopic enrichment	75% Nuclebras 15% Interatom 10% Steag Kemenergie (FRG)
Nuclemon	Heavy minerals	100% Nuclebrás
Nuclen	Engineering	75% Nuclebrás 25% KWU (Kraftwerk Union) (FRG)
Nuclep	Heavy components	97.6% Nuclebrás 0.8% KWU 0.8% Voest-Alpine (Austria) 0.8% GHH
Nucon	Construction	100% Nuclebrás
Nustep *	Isotopic enrichment	50% Nuclebrás 50% Steag

Source: Nuclebrás, *Annual Report*, 1984.
*Located in FRG.

technology was not commercially proven, and its electricity consumption was high. The scientific community backed a more independent approach, possibly along the natural uranium, heavy water lines (such as the Argentine program), stressing the advantages of a thorium-based cycle, because Brazil had considerable reserves of thorium, as well as an incipient national technology. Domestic entrepreneurs in the capital goods and engineering sectors were far from satisfied with the role they were allocated in the joint ventures with the German partners. In its political, economic, and technical nature, the arrangement expressed the encroachment of the economic ministries and their supremacy in shaping sectoral policies in tune with the Brazilian regime's broad macropolitical objectives.[52]

The exorbitant rise in the cost of the planned nuclear program, from an original estimate of about $10 billion in 1975 to a projected $40 billion in the early 1980s, generated widespread criticism.[53] Compounded by disclosures of mismanagement and corruption, mounting economic difficulties, poor selection of sites, and lack of adequately trained personnel, the implementation of the 1975 agreement was delayed and finally sharply contracted in the mid-1980s. Criticism also focused on KWU's poor record in technology transfer, leading to the resignation of two directors of Nuclebrás subsidiaries.[54] General Dirceu Lacerda Coutinho, director of Nuclei (isotopic enrichment), forwarded a critical report to the National Security Council, the National Information Services, and to President Geisel, without reply, and later testified in a congressional investigating committee on German reluctance to transfer technology effectively.[55]

By 1988, the accomplishments of the 1975 nuclear agreement could be summarized in the inauguration of (an idle) heavy components factory (Nuclep), a fuel element fabrication plant, and two unfinished power stations. A uranium concentrate plant designed by the French company Pechiney Ugine Kuhlman was inaugurated in 1982. Following the May 1980 Argentine-Brazilian nuclear cooperation agreements, Argentina commissioned Nuclep to weld and assemble the lower part of the pressure vessel for its Atucha II plant. Through its partnership with KWU, Nuclep participated in bidding for the supply of two reactors to Mexico.[56] Brazil also signed an agreement in 1980 to provide Iraq with natural and low-enriched uranium, equipment, personnel training, and technology for reactor construction, and it reportedly shipped 240 tons of uranium to Iraq.[57] Brazil's export potential, however, ultimately may be seriously constrained by limited global demand and a thriving competition.[58]

Critics of the nuclear program within the technocracy rallied around General Costa Cavalcanti, director of Eletrobrás and Itaipu Binational, who advocated the development of Brazil's hydroelectric potential.[59] With the appointment of Antônio Delfim Netto, one of the architects of the liberal economic policy, as minister of planning, this group succeeded in reversing priorities in favor of hydroelectricity in the late 1970s. The decline of Nuclebrás and the official program with KWU can be contrasted with the rise of the National Nuclear Energy Commission under National Security Council protection. The commission headed a "parallel program," invigorated in the early 1980s, designed to advance indigenous nuclear technology.[60] Among its achievements was the inauguration in 1988 of a uranium enrichment facility under navy control. Its projects, unlike those of the German agreement, are not covered by international safeguards that would prevent Brazil from diverting fuel or replicating technologies to obtain weapons-grade materials. The Brazilian air force has a nuclear research institute in the Centro Técnico Aeroespacial in São José dos Campos, and the army reportedly concentrates on the use of nuclear energy for satellite propulsion in its own large research center, the

Centro Tecnológico do Exército (Centex).[61] In September 1983, Navy Minister Admiral Maximiano da Fonseca announced that Brazil would begin construction of its first atomic submarine in the early 1990s.[62]

Most analyses of the cluster of incentives that might drive Brazil to acquire a nuclear capability have traditionally focused on issues of international status, prestige, and independence, particularly independence from the United States.[63] Neither the "pariah" nor the "pygmy" characteristics of other countries on the nuclear threshold apply to Brazil.[64] Its declaratory policy points to the denial of nuclear technology to nonsignatories of the Non-Proliferation Treaty as part of a general attempt by the nuclear powers to perpetuate international stratification. At least some Brazilians may regard an independent nuclear weapons capability as a useful diplomatic tool to increase both Brazil's stature among developing countries and its leverage vis-à-vis the developed world. Such capability would portend, in this view, an upgrading of Brazil's credentials as a major contender in the international and regional arenas. However, prestige as an incentive may be offset by the need to come to terms with the severe socioeconomic effects of many of the grandiose schemes undertaken in the 1970s. Economic and political viability, not *grandeza* (greatness), is the adjusted objective of the late 1980s. Moreover, there is growing understanding among nuclear-capable Third World countries that an overt military nuclear capability may be of decreasing strategic value. Once capabilities are in place, intentions remain under a cloud of ambiguity, as in the case of Pakistan—broadening the repertoire of nuclear postures as an instrument of foreign policy.

Few studies agree on the impact of regional security considerations, more specifically the Argentine factor, on Brazilian incentives.[65] In the aftermath of Argentine defeat in the Malvinas/Falklands War, arguments citing an Argentine effort to regain lost capabilities acquired greater currency in Brazil. Yet the significance of this factor seems to have been overstated in light of the relatively benign security environment of the Southern Cone of South America, especially in comparison with other regions such as the Middle East and South Asia. The series of agreements on nuclear technology cooperation between Argentina and Brazil since 1980 has led to presidential-level mutual visits to sensitive facilities.[66] Brazil supplied Argentina with the pressure vessel and steam generators for the Atucha II plant, and Argentina supplied Brazil with zircaloy tubing and lent it 240 tons of uranium for Angra I, part of which was returned after production started in Poços de Caldas.

The military establishments in both Brazil and Argentina are often cited for efforts to accelerate their respective nuclear programs, but in recent years the levels of antagonism have been low. The Brazilian armed forces control strategic technical areas including nuclear technology, telecommunications, weapons, aeronautics, and computers.[67] However, the military is no monolithic entity, and there are growing ideological cleavages concerning broad economic policies, state control, global alliances, and the objectives of the nuclear program. The foreign policy of caution and restraint on the part of former Argentine President Alfonsin helped diffuse concerns among the Brazilian military.

Whether the bilateral agreements of the 1980s (including joint development of a breeder reactor) reflect tactical cooperation geared to oppose foreign pressures on their nuclear programs or signs of sincere rapprochement, the two countries have set in motion a process of mutual accommodation that reduces the proliferation incentives linked to regional competition. Like economic interdependence and cooperation, prestige derived from nuclear competence can be seen as a positive sum game. Competition between Argentina and Brazil is more likely to occur at the level of nuclear technology exports to other less developed countries, particularly in Latin America. However, cooperation in the form of joint ventures with third partners, including possibly Cuba, is also feasible.[68]

Nuclear power may not be the most efficient solution to Brazil's energy problems, at least in the short run. Economic and sociopolitical realities impose severe constraints on capital-intensive nuclear-related activities, peaceful and otherwise. Yet, in the words of a Brazilian colonel, "A nuclear program is fundamentally, and almost exclusively, a matter of national security. The harnessing of energy from a nuclear reactor is secondary."[69] Brazil has not signed the Non-Proliferation Treaty (NPT) and there is broad domestic consensus, even among critics of the nuclear program, that vertical proliferation (additions to superpowers' arsenals) poses a much greater threat to humanity than horizontal proliferation (growth in the number of nuclear powers); in this view, the NPT is designed primarily to perpetuate a discriminatory distribution of power.

In 1962, under a civilian regime, Brazil proposed to the United Nations the creation of a nuclear-free zone in Latin America, which eventually resulted in the Treaty of Tlatelolco. In 1978, Brazil ratified the treaty but chose not to waive the requirements of article 28, paragraph I, and in practical terms the treaty is therefore not in force as far as Brazil is concerned.[70] From the legal standpoint, once the treaty comes into force for a country, it is obliged to negotiate a full-scope safeguards agreement with the International Atomic Energy Agency. Brazil perceives the treaty as not impinging on the signatories' right to conduct peaceful nuclear explosions.[71] The safeguard provisions of the 1975 West German agreement were consistent with the guidelines of the Zangger Committee on nuclear exports at that time. Brazil did not, however, accept full-scope safeguards covering its entire nuclear industry. In practical terms, moreover, the agreement with West Germany allowed room for movement of nuclear materials in and out of safeguarded facilities, under article VII.

Brazil's position with respect to the NPT and Tlatelolco, its refusal to agree to full-scope safeguards, and its insistence on its right to conduct peaceful nuclear explosions point to a policy stressing national autonomy and strategic flexibility.[72] Cooperation through agreements on hydro-electrical and other natural resources development, including nuclear technology, have served Brazilian interests better than an arms race, at both global and regional levels. According to one analysis of military competition in South America, Brazil's military expenditures—about 1 percent of its GDP—are the only ones in the region compatible with peacetime growth.[73]

Brazil, the NICs, and Adjustment Patterns

During hard times, argues Peter A. Gourevitch, "patterns unravel, economic models come into conflict, and policy prescriptions diverge."[74] Brazil's responses to the oil shocks of the 1970s contributed to the expansion of its state enterprises, the deepening of its foreign financial indebtedness, the intensification of countertrade strategies, and the sobering of the ruling coalition's own evaluation of its macropolitical model of economic change. Some of Brazil's solutions increased its external vulnerability and fomented friction with environmental and nonproliferation groups in North and South alike; the domestic social costs were even higher. If anything, the energy crisis highlighted, for Brazil as for others, the interdependence of politics and economics, and of global and domestic, in the definition of states' response to external challenges.

Brazil was only one among those affected by the oil crises, which helped truncate the rapid economic advancement of the pre-1973 period. In the absence of a multilateral challenge to OPEC, Brazil's ruling coalition chose to rely primarily on its own efforts at domestic restructuring. The emphasis on domestic strategies of adjustment was largely molded by Brazilian decision-makers' perception that the state's capacity to internalize the costs of adjustment was greater than its international ability to help shape a new regime via, for instance, mobilizing support against OPEC among other Third World countries.[75] Brazil's diplomacy was thus characterized by a pragmatic and technocratic bargaining style, in which fluid policies tended to rely on bilateral relations as being more dependable than multilateral cooperation.

Beyond the domestic rationale advanced in this paper, three system-level types of mutually reinforcing explanations can be put forward to explain Brazil's selection of instruments of adjustment to the external challenges of energy markets. First, its mild reactions may be seen as buttressing the contention that developing countries' support for common political objectives, such as global restructuring of international regimes (in this case by Third World oil producers), can override concerns with their own economic performance.[76] In its response, Brazil tacitly yielded to a transformed energy market without pressing for the creation of an authoritative international regime that would render predictable, if not secure, resource transfers to affected developing countries.

A second explanation may be found in the country's structural position in the international political economy. As part of what is at times labeled the "semiperiphery," Brazil is caught in the ambiguity that the advantages of an upgraded international status may create.[77] Although not yet reaping the benefits of being a full-fledged member of the advanced tier, its interests increasingly diverge from those of less developed countries left behind. It expects to maintain smooth ties with the industrialized world, which supplies it with the capital, technology, and export markets on which its growth is dependent, while securing Third World export markets and sources of raw materials.

Finally, Brazil's response to the energy debacles of 1973 and 1979—not unlike those of other oil importers in the industrialized world—can be interpreted as reflecting an inexorable shift of the international system from a "military" to a trading world.[78] Diplomacy and reciprocity, financial flows, domestic adjustments, and export drives prevailed over more belli-cose alternatives.

Brazil's energy predicament—which turned it into the Third World's largest oil importer—may be compared to that of other energy-poor developing countries with nuclear programs reviewed in this book. These countries' relative vulnerability to changes in oil prices and supplies is linked to their respective resource endowments, degrees of energy depen-dence, and patterns of economic development.[79] Less Developed Coun-tries undergoing rapid structural change and increasing energy consump-tion are particularly vulnerable. In light of their rising position in the international economic system and their energy dependence, they face similar challenges, leading, at times, to similar responses. For instance, both India and Brazil changed their policies regarding oil exploration and self-reliance by inviting foreign companies to participate, but both failed in attracting OPEC investments. Yet vulnerability to foreign oil supplies does not seem to be associated with uniform patterns of state intervention and entrepreneurship. Thus, the expansion of state firms in Brazil in the 1970s can be contrasted with the market-conforming strategies of South Korea, which allowed private firms greater opportunities in energy markets.

Most East Asian nonindustrialized countries (NICs), including South Korea, Taiwan, and Singapore, were struck by the quadrupling of oil prices in 1973, but the domestic costs of adjustment were lower for their smaller, flexible, trading economies and their relatively more egalitarian social structures. Their responses leaned more on fundamental economic restructuring and acceptance of lower initial growth to limit imports while promoting exports to allow economic recovery, than, as in the case of Brazil, on foreign indebtedness.[80] Despite these instrumental differences, however, most NICs responded by maintaining centralized authoritarian political models backed by military force. Their ability to impose costs on domestic groups may decrease with political liberalization, compelling democratic regimes to search for a different mix of domestic and international—including multilateral—solutions. A particularly constraining factor for Latin American NICs has been their external indebtedness, which increases their vulnerability to external leverage in shaping political and economic objectives.

Beyond their common concern with energy considerations, NICs vary not only with respect to the vulnerability of their economic infrastructures and the state's ability to steer investment patterns, but also in the levels of regional security threats. In South America, regional power politics and deterrence are far less relevant than are other international considerations. The security factor sets Brazil apart from India, Israel, Taiwan, Pakistan, South Korea, and South Africa. There has been little serious probability of superpower entanglement in the Southern Cone, even in the era before glasnost, which reduced even further the incentives for regional military competition.

The availability of alternative sources of energy in Brazil, as in Argentina and India, seriously calls into question the major commitments to nuclear power by at least some of the NICs. The decisions to embark on such programs were associated more with "pull" (domestic considerations) than "push" factors (suppliers' intervention). The programs' political, economic, and technical characteristics were shaped by the countries' respective domestic structural and institutional arrangements. The avenue of acquiring a nuclear weapons option through a civilian power program, as the case of India appears to suggest, inextricably links questions of security with issues of economic growth and technological development. In most cases, nuclear power has been seen as a litmus test of independence, sophistication, and industrial technological advancement. Yet the contribution of nuclear power to the broader modernization of these countries' industrial structure has been limited in most cases, while the nuclear sector has absorbed resources far greater than the economic benefits it could provide.[81]

Diverging trade, financial, and strategic considerations have placed NICs, in some areas at least, at opposing ends of multilateral bargaining processes. Domestic structures and policy networks shaped different state capacities and proclivities. These two factors, in turn, defined the balance of incentives and constraints associated with different strategies, and ultimately influenced the nature of a country's adjustments to increased interdependence. Changing economic and technological capabilities for the group as a whole may reduce external vulnerabilities with regard to energy, while global changes—notably the decline of east-West tensions—may provide new opportunities for expansion of their economic and political relations, ignoring ideological considerations in much the pragmatic style adopted by Brazil during the first oil shock in 1973.

NOTES

I would like to acknowledge the assistance of UCLA's Center for International and Strategic Affairs and Karl Arnold in particular as well as the University of California's Institute on Global Conflict and Cooperation. I also thank Ken Conka, Jan K. Black, Jim Dietz, Kenneth Erickson, Patrice F. Jones, Pat Peritore, Wayne Selcher, Scott Tollefson, Steven Topik, and Maria Jose Willumsen for their constructive comments.

1. Such integration occurred not exclusively via exports but also through increased imports of capital goods, intermediate inputs, technology, capital, and foreign investment. Albert Fishlow, "Latin American Adjustment to the Oil Shocks of 1973 and 1979," in J. Hartlyn and S.A. Morley, eds., *Latin American Political Economy* (Boulder: Westview Press, 1986), pp. 54-84. Celso Lafer, Política Exterior Brasileira: Balanço e Perspectivas," *Dados* 22 (1979): pp. 49-64.

2. Petroleum imports were no higher than 12 percent of total imports until 1973. On the draining effects of oil imports, see James H. Street; "Coping with Energy Shocks in Latin America: Three Responses," *Latin America Research Review* 17, no. 3 (1982): pp. 128-47; Albert Fishlow, "A Tale of Two Presidents: The Political Economy of Crisis Management," and E. Bacha and P. Malan, "Brazil's Debt: From the Miracle to the Fund," in A. Stepan, ed., *Democratizing Brazil* (New York: Oxford University Press, 1989), pp. 83-119.

3. William H. Courtney, "Nuclear Choices for Friendly Rivals," in Joseph A. Yager, *Nonproliferation and US Foreign Policy* (Washington, D.C.: Brookings Institution, 1980).

4. Fishlow, "Latin American Adjustment." The average annual rate was close to 6.5 percent in this period.

5. In other words, did the state attempt to control energy companies or to

strengthen them without direct intervention? I have adopted these categories from Richard S. Samuels, *The Business of the Japanese State—Energy Markets in Corporative and Historical Perspectives* (Ithaca: Cornell University Press, 1987).

6. W. Baer, *The Brazilian Economy* (New York: Praeger, 1983). For a comprehensive view of energy in Brazil, see Kenneth P. Erickson's "The Energy Profile of Brazil," in Kenneth R. Stunkel, ed., *National Energy Profiles* (New York: Praeger, 1981), and his "State Entrepreneurship, Energy Policy, and the Political Order in Brazil," in T.C. Bruneau and P. Faucher, eds., *Authoritarian Capitalism: Brazil's Contemporary Economic and Political Development* (Boulder: Westview Press, 1981), pp. 141-77.

7. Shares in net assets of petroleum refining and distribution were 4 percent for domestic firms, 12 percent for foreign ones, and 84 percent for state companies in 1981.

8. Petrobrás accounts for close to 40 percent of local demand for capital goods (*Brasil Energia*, February 1984). For a comprehensive study of the formative years of Petrobrás and their legacy, see John D. Wirth, ed., *Latin American Oil Companies and the Politics of Energy* (Lincoln: University of Nebraska Press, 1985); Peter S. Smith, *Oil and Politics in Modern Brazil* (Toronto: Macmillan of Canada, 1976); and George Philips, *Oil and Politics in Latin America—Nationalist Movements and State Companies* (Cambridge: Cambridge University Press, 1982).

9. *A Questão Nuclear-Aspectos Conjunturais de Energia*, Brasilia Senado Federal, 1983. See also *Indice do Brasil, 1979-1980*, O Banco de Dados, 1981, p. 36, and *New York Times*, March 3, 1989. For a landmark study of the hydroelectric sector in Brazil, see Judith Tendler, *Electric Power in Brazil: Entrepreneurship in the Public Sector* (Cambridge, Mass.: Harvard University Press, 1968).

10. The World Bank in fact withheld a $500 million loan for the energy sector to await environmental safeguards.

11. Thomas J. Trebat, *Brazil's State-Owned Enterprises—A Case Study of the State as Entrepreneur* (Cambridge: Cambridge University Press, 1983).

12. There are plans to build seventy new dams in the Amazon Basin and to transfer electricity from there to the energy-intensive areas of the southeast.

13. Diesel and gasoline for transportation account for over 50 percent of total oil consumption.

14. The Proalcohol program, created in 1975, saved about $4.5 billion in crude oil imports and created 148,000 new jobs, most of them in the agricultural sector, in 1980 alone (*Monthly Letter*, Banco do Brasil, no. 49 [January 1983]: p. 2; *Brazil Energy*, July 28, 1982, p. 5). However, its social costs were quite high, particularly in the displacement of agricultural labor in favor of capital-intensive sugar cane.

15. Michael Barzelay, *The Politicized Market Economy* (Berkeley: University of California Press, 1986).

16. Coal mining started in Brazil in 1966; the increase in production has been minimal. Affonso da Silva Telles, "Participação do carvão na produção de energia elétrica—estudo para planejamento," in L. Pinquelli Rosa, ed., *Energia, Tecnologia e Desenvolvimento* (Petrópolis: Vozes, 1978).

17. Fishlow, "Latin American Adjustment." Rising oil prices and mounting interest rates accounted for over half of the $4 billion deterioration in the current account in 1979.

18. Erickson, "The Energy Profile," p. 231.

19. *Brazilian Monthly Economic Indicators*, Planning Secretariat of the Presidency, Federative Republic of Brazil, January 1983, May 1983, January 1984.; *Monthly Newsletter*, Central Bank of Brazil, October 1988.

20. In fact, the trade deficit with the Middle East more than doubled between 1986 and 1987. *Monthly Newsletter*, February 1988.

21. Third and fourth, respectively, of the eight originally included in the agreement with West Germany.

22. GNP growth does not necessarily depend on exponential growth in energy consumption. Nazli Choucri, *Energy and Development in Latin America* (Lexington, Mass.: Lexington Books, 1982), p. 10.

23. The latter were given four years to adjust to new regulations, which tend to be more flexible concerning raw materials processed in Brazil.

24. Trebat, *Brazil's State-Owned Enterprises*. A $500 million loan to Eletrobrás approved by the World Bank in 1988 has been delayed because of the transfer of nuclear plants to Eletrobrás jurisdiction (*J.P.R.S.* Jan. 13, 1989).

25. *O Estado de São Paulo*, Oct. 22 and Dec. 21, 1985. Thirty percent of the $9 billion in direct investment in the Itaipu hydroelectric project was financed through external debt. Brazil has been paying $1 billion per year in interest and other charges related to the nuclear sector. *Correio Brasiliense*, July 16, 1986; *New York Times*, Dec. 21, 1987.

26. G. John Ikenberry, "The Irony of State Strength: Comparative Responses to the Oil Shocks in the 1970s," *International Organization* 40, no. 1 (Winter 1986): pp. 105-37. Samuels, *The Business of the Japanese State*, p. 18; Harvey B. Feigenbaum, *The Politics of Public Enterprise: Oil and the French State* (Princeton: Princeton University Press, 1985).

27. A.F.K. Organski, *World Politics*, 2d ed. (New York: Knopf, 1986).

28. U.S. Arms Control and Disarmament Agency, *World Military Expenditures and Arms Transfers, 1967-76*, (Washington D.C., 1978).

29. Brazil, India, Taiwan, South Korea, Egypt, and Mexico accounted for 75 percent of the total rise in LDC oil consumption between 1974-78; P. Kemezis and E.J. Wilson III, *The Decade of Energy Policy* (New York: Praeger, 1984) p. 54.

30. Wayne A. Selcher, *Brazil in the Global Power System*, Occasional Papers Series, Center of Brazilian Studies, Johns Hopkins University, November 1979. For other authoritative studies of Brazil's foreign policy, see Riordan Roett, ed., *Brazil in the Seventies* (Washington, D.C.: American Enterprise Institute for Public Policy Research, 1976) and Ronald M. Schneider, *Brazil—Foreign Policy of a Future World Power* (Boulder: Westview Press, 1976).

31. Wayne A. Selcher, *Brazil's Multilateral Relations between First and Third Worlds* (Boulder: Westview Press, 1978), p. 14.

32. Iraq is said to have paid about $1.2 billion between 1976 and 1981 for Brazilian military equipment (*Los Angeles Times*, Nov. 15, 1981). According to Scott Tollefson, reported estimates of $2 billion to $5 billion in annual arms exports for the early 1980s are exaggerated ("Brazilian Arms Sales and Foreign Policy: The Search for Autonomy," Ph. D. Dissertation, Johns Hopkins University, forthcoming).

33. *Brazil Energy*, July 28, 1982. See also Tollefson, "Brazilian Arms Sales."

34. *Monthly Newsletter*, Central Bank of Brazil, September 1988. Over 30 percent of Brazil's total imports came from the Middle East by 1981, as opposed to only 5 percent in 1970. See Baer, *The Brazilian Economy*, p. 163.

35. Philips, *Oil and Politics*.

36. *Latin America Political Report* 12, no. 22 (1978): p. 4.

37. John Child, "Geopolitical Thinking in Latin America," *Latin America Research Review* 14, no. 2 (1979).

38. M. Travassos, *Projecao Continental do Brasil* (Sao Paulo: Cia. Editora Nacional, 1947).

39. This statement applies less to the early 1970s—particularly with regard to Argentina, Uruguay, Bolivia, and Chile—than to the post-1974 era. The hegemonic

aspects of Brazilian foreign policy are stressed in Juan E. Guglialmelli, *Argentina, Brasil y la bomba atómica* (Buenos Aires: Tierra Nueva, 1976); William Perry, *Contemporary Brazilian Foreign Policy: The International Strategy of an Emerging Power* (Beverly Hills, Calif.: Sage Publications, 1976); Phillip Kelly, "Geopolitical Tension Areas in South America: The Question of Brazilian Territorial Expansion," in R.E. Biles, ed., *Inter-American Relations: The Latin American Perspective* (Boulder: Lynne Rienner, 1988); and J. Child, *Geopolitics and Conflict in South America: Quarrels among Neighbors* (New York: Praeger, 1985). On Brazil's reluctance to assume a hegemonic role in South America, see Wayne A. Selcher, "Brazil and the Southern Cone Subsystem," in Pope Atkins, ed., *South America in the 1990s: Evolving International Relationships in a New Era* (Boulder: Westview Press, forthcoming), and William Perry, "Brazil: A Local Leviathan," in Rodney Jones and S. Hildreth, eds., *Emerging Powers: Defense and Security in the Third World* (New York: Praeger, 1986).

40. Thomas E. Skidmore, *The Politics of Military Rule in Brazil, 1968-85* (New York: Oxford University Press, 1988); David J. Myers, "Brazil: Reluctant Pursuit of the Nuclear Option," *Orbis*, Winter 1984: pp. 881-911.

41. Howard Pittman, "Geopolitics and Foreign Policy in Argentina, Brazil and Chile," in E.G. Ferris and J.K. Lincoln, eds., *Latin American Foreign Policies: Global and Regional Dimensions* (Boulder: Westview Press, 1981).

42. The conflict over the Banda Oriental resulted in the creation of Uruguay as a buffer state. In the 1932-35 Chaco War between Bolivia and Paraguay, Brazil and Argentina supported opposing sides struggling for control over territory of presumed economic importance.

43. Hydroelectric projects are seen not merely as an answer to national energy requirements but as a developmental goal in themselves, due to their presumed infrastructural, industrial, and other socioeconomic spinoffs.

44. The completion of Itaipu in early 1983, however, sealed off the higher Paraná from navigation southward, diminishing the economic importance of the Argentine section of the river and reducing the electricity production capability of the Argentine Corpus dam.

45. Margaret Daly Hayes, *Brazil and the South Atlantic: Perspectives on an Emerging Issue*, Occasional Papers Series, Center of Brazilian Studies, Johns Hopkins University, n.d.

46. Riordan Roett, "The Political Future of Brazil," in William H. Overholt, ed., *The Future of Brazil* (Boulder: Westview Press, 1978), pp. 71-102; Thomas E. Skidmore, "Brazil's Changing Role in the International System: Implications for U.S. policy," in Roett, *Brazil in the Seventies*.

47. A series of accords signed by Secretary of State George Shultz in Brazil in February 1984 approved U.S. participation in a $4 billion hydroelectrical project and a $50 million helicopter coproduction deal; *Los Angeles Times*, Feb. 7, 1984. On tensions regarding nuclear policy, trade, and human rights, see Robert Wesson, *The United States and Brazil* (New York: Praeger, 1981) and Albert Fishlow, "The United States and Brazil: The Case of the Missing Relationship," *Foreign Affairs* 60, no. 4 (Spring 1982): pp. 904-23.

48. Wayne A. Selcher, "Brazil in the World: Multipolarity as Seen by a Peripheral ADC Middle Power," in E.G. Ferris and J.K. Lincoln, eds., *Latin American Foreign Policies: Global and Regional Dimensions* (Boulder: Westview Press, 1981), pp. 81-102.

49. The Instituto de Energia Atomica (São Paulo), Instituto de Pesquisas Radioativas (Belo Horizonte), and Comitê Nacional de Energia Atomica (CNEN; Rio de Janeiro) were created in this period.

50. West Germany would receive 20 percent initially of any ore discovered and a larger share at a later date.

51. Brazilian negotiators argued also that reprocessing and enrichment would lead to considerable savings in uranium consumption and to a reduction in waste-related problems. It would also provide the knowledge required to obtain weapons-grade fuel; *Jornal do Brasil*, Feb. 17, 1979. José Goldemberg, *Energia nuclear no Brasil* (São Paulo: Hucitec, 1978); Joaquim F. de Carvalho, *O Brasil Nuclear* (Porto Alegre: Tché!, 1987).

52. Etel Solingen, *Bargaining in Technology: Nuclear Industries in Brazil and Argentina* (in preparation).

53. By 1980, the cost of the second and third plants (Angra II and III) was calculated at $3.1 billion each, over twice the original estimate. The costs of a nuclear-generated kilowatt shot up from $200 in 1970 to $2,800 in 1983. A former director of Nuclebrás engineering subsidiary Nuclen, Joaquim de Carvalho, claimed that by the late 1970s the program was no longer geared to the production of economically priced energy but to building plants "at whatever cost." *Brazil Energy*, July 10, 1980; March 24, 1981; October 10, 1981.

54. César Cals, minister of mines and energy, declared in 1971 that "nuclear plants do not obey energy criteria but criteria of absorption of technology"; *Jornal do Brasil*, Aug. 8, 1981. Yet Joaquim de Carvalho, former director of Nuclen (Nuclebrás Engineering), claimed that Nuclen serves only to coordinate the production of German-designed components and supervise their assembly—"nothing more"; *Brazil Energy*, March 24, 1980.

55. *Worldwide Report—Nuclear Development and Proliferation*, F.B.I.S. no. 193 (June 24, 1983): pp. 14-15. The poor performance of Westinghouse's Angra I exacerbated sensitivities regarding effective technology transfer. After twelve years of delay, Brazil's first, has never been fully operational due to technical deficiencies that brought the utility Furnas to file a suit against Westinghouse.

56. *Brazil Energy*, Dec. 24, 1981.

57. *Brazil Energy*, June 24, 1981.

58. For a full description of Brazil's nuclear industrial capabilities, including export potential, see Etel Solingen, "Technology, Countertrade and Nuclear Exports," in W.C. Potter, ed., *International Nuclear Trade: The Challenge of the Emerging Suppliers* (Lexington, Mass.: Lexington Books, 1990).

59. Capital costs represent 70 percent of a power station. In Brazil, the generating costs of nuclear facilities were between two and three times those of the hydroelectric sector; Luiz Pinguelli Rosa, *Nuclear Energy in Latin America: The Brazilian Case*, U.N. University, 1980.

60. *Latin America Weekly Report*, Feb. 11, 1983. According to Myers, "Brazil: Reluctant Pursuit," and F.B.I.S. reports, CNEN director Rex N. Alves, the liaison with the National Security Council, was reportedly the man behind the delivery of uranium to Iraq.

61. *Latin America Regional Reporters*, Brazil, Feb. 5, 1982.

62. *O Estado de São Paulo*, Sept. 24, 1983.

63. A long-standing effort to increase autonomy was reinforced by the American violation in 1974 of a 1972 agreement to supply enriched uranium for the Westinghouse plant.

64. Richard Betts, "Paranoids, Pygmies, Pariahs and Nonproliferation," *Foreign Policy* no. 26 (Spring 1977): pp. 157-83.

65. Ernest W. Lefever, *Nuclear Arms in the Third World* (Washington, D.C.: Brookings Institution, 1979); George H. Quester, *Brazil and Latin American Nuclear*

Proliferation: An Optimistic View, UCLA Center for International and Strategic Affairs, ACIS Working Paper 17, 1979; Courtney, "Nuclear Choices"; John H. Rosenbaum, "Brazil's Nuclear Aspirations," in Onkar Marwah and Ann Schulz, eds., *Nuclear Proliferation and the Near Nuclear Countries* (Cambridge, Mass.: Ballinger, 1975), pp. 255-77. For an alternative view of the impact of the regional system on proliferation, see Ashok Kapur, "The Proliferation Factor in South America: the Brazil-Argentine Cases," in *International Nuclear Proliferation—Multilateral Diplomacy and Regional Aspects* (New York: Praeger, 1979); and Regina Lucia de Moraes Morel, *Ciencia e Estado—A política Científica no Brasil* (São Paulo: T.A. Queiroz, 1979).

66. The meeting between Presidents Figueiredo and Jorge R. Videla, the first visit to Argentina by a Brazilian president since 1945, was considered a turning point in the relationship between the two countries. It resulted in a series of cooperative agreements on hydroelectricity, natural gas, alcohol, and nuclear technology.

67. Simon Schwartzman, *Ciencia, Universidade e Ideologia—A Política do Conhecimento* (Rio de Janeiro: Zahar, 1981).

68. After 1979, Brazil accelerated its nuclear cooperation program with Venezuela and Chile in an attempt to trade nuclear technology for oil and utilize Nuclep's idle capacity.

69. Erickson, "The Energy Profile."

70. Until certain requirements are fulfilled, namely, ratification by all Latin American nations (Argentina and Cuba have not yet ratified the treaty). John R. Redick, "Nuclear Proliferation in Latin America," in Roger W. Fontaine and James D. Theberge, eds., *Latin America's New Internationalism—The End of Hemispheric Isolation* (New York: Praeger, 1976), and "The Tlatelolco Regime and Nonproliferation in Latin America," in George H. Quester, ed., *Nuclear Proliferation—Breaking the Chain* (Madison: University of Wisconsin Press, 1981): pp. 103-34.

71. The contention has been that the latter are important in developmental projects such as the extraction of oil from shale; the linking of the Plata, Amazon, and Orinoco rivers to integrate South America; and the construction of dams and canals. Even in 1967, the Brazilian interpretation of Tlatelolco's article 18 was that it allows signatories to undertake "by their own means or in association with third parties, nuclear explosions with peaceful purposes including those presupposing devices similar to the ones utilized in military weapons"; Kapur, "Proliferation Factor."

72. Courtney, "Nuclear Choices for Friendly Rivals."

73. Emilio Meneses, "Competencia Armamentista en América del Sur 1970-1980," *Estudios Públicos* (Santiago), no. 7 (1982): pp. 5-42.

74. Peter A. Gourevitch, *Politics in Hard Times—Comparative Responses to International Economic Crises* (Ithaca: Cornell University Press, 1986).

75. In this sense, Brazil's reactions closely resemble those of statist models like France, in its emphasis on national control over energy production and supply, state-induced commercial and barter agreements with suppliers, and the adoption of an ambitious nuclear energy program; G. John Ikenberry, *Reasons of State: Oil Politics and the Capacities of American Government* (Ithaca: Cornell University Press, 1988).

76. Stephen D. Krasner, *Structural Conflict—The Third World Against Global Liberalism* (Berkeley: University of California Press, 1985).

77. Peter Evans, *Dependent Development—The Alliance of Multinational, State and Local Capital in Brazil* (Princeton: Princeton University Press, 1979); Immanuel Wallerstein, "Semi-peripheral Countries and the Contemporary World Crisis," *Theory and Society*, 3, no. 4 (1976): pp. 461-84.

78. Richard Rosecrance, *The Rise of the Trading State* (New York: Basic Books, 1986).

79. Joy Dunkerley et al., *Energy Strategies for Developing Nations* (Baltimore: Johns Hopkins University Press, 1981).

80. Fishlow, "Latin American Adjustment." South Korea's foreign indebtedness, however, grew considerably as well.

81. Solingen, *Bargaining in Technology.*

8

Cuba

JORGE F. PÉREZ-LÓPEZ

In June 1960, the Cuban revolutionary government seized the refineries operated by three international oil companies and severed all relationships with its traditional energy suppliers. The economic collapse that would have been predictable in imported energy–dependent Cuba as a result of these events was avoided by swift action of the Soviet Union in setting up an "oil bridge" to supply Cuba from Black Sea ports.

More than two decades later, Cuba's dependence on the Soviet Union for energy products continues and, in effect, has deepened. Although Cuban oil production has increased significantly in the 1980s, it continues to account for a very limited percentage of consumption. Reliance on a single supplier for the overwhelming majority of energy imports makes Cuba vulnerable to Soviet supply disruptions or manipulations. Nuclear power, which could ease dependence on hydrocarbon imports, will have little impact on Cuban vulnerability, as the only source of nuclear hardware and fissionable materials is the Soviet Union.

This paper explores the relationships between energy, security, and the economy in revolutionary Cuba. In the first section a rough balance of energy production, imports, and consumption is developed.[1] The next section explores the national security implications of Cuba's dependence on imported fuels and discusses the impact of such dependence on foreign policy, drawing on a factual example of oil supplies allegedly being withheld until the Cuban government made certain policy changes. The third section analyzes the role of energy in the economy, reviews the special relationship between Cuba and the Soviet Union regarding oil pricing, and speculates on Cuba's energy future in view of uncertainties about the willingness of the Soviet Union to continue to supply the country with growing volumes of oil and oil products.

As a result of its topography and geology, Cuba is poorly endowed with energy resources. The narrow, elongated shape of the island precludes the existence of large water masses capable of producing hydroelectricity. Despite considerable exploration, the deposits of hydrocarbons that have been found are grossly inadequate when compared with total needs.

Imported oil and oil products, and to a lesser extent coal, have historically filled the gap between energy demand and domestic production. Nuclear power will have no effect on the energy balance until the early 1990s at the earliest.

DOMESTIC ENERGY PRODUCTION

Domestic sources of primary energy in Cuba during 1959-87 for which data or estimates are available include oil, natural gas, hydroelectricity, ethanol, bagasse, fuelwood, and charcoal. For commercial fuels (oil, natural gas, hydroelectricity, and ethanol), published annual production data appear to be good indicators of domestic availability. This is not the case for noncommercial fuels (bagasse, fuelwood, and charcoal), which are often consumed directly by the producers and thus do not register in the statistical system. For bagasse, production data are not available but can be estimated under some limiting assumptions; for fuelwood and charcoal, published data probably underestimate actual production by a substantial margin.

Commercial oil production in Cuba began in 1915 with the discovery of the Bacuranao field; another commercial field was discovered at Jarahueca in 1943. Output from these two fields was small, averaging about 4,000 metric tons (MT) per year during 1950-54. The discovery of the important Jatibonico field in 1954 pushed production to an average of about 30,000 MT per year during 1955-58 and gave rise to a flurry of concession applications and exploratory drilling activities by domestic and foreign companies. Small fields were discovered subsequently at Catalina, Cristales, and Guanabo. However, as most of the exploratory wells either turned up dry or found petroleum in quantities too small or too low in quality to justify commercial exploitation, the exploration boom subsided.

The revolutionary leadership in 1959 was convinced that Cuba had vast oil reserves not exploited by foreign oil companies operating there, since the companies could reap higher profits from refining and marketing imported crude. At the end of October 1959, the Cuban government seized the exploration records of the oil companies and, with financial and technical assistance from the Soviet Union and Romania, undertook an ambitious program aimed at boosting oil production.

Initially, the program had very limited success. Oil production for 1960-67 averaged 50,000 MT per year, exceeded 200,000 MT per year in 1968-69 when output peaked at the Guanabo field, then steadied at about 140,000 MT per year during 1970-74, as production declined at mature fields (table 8.1). Production gains from the newly discovered Boca Jaruco and Varadero fields east of Havana pushed output above 260,000 MT per

Table 8.1.   Cuban production of commercial energy, 1959-1987

	Oil (000 MT)	Natural gas (million cubic meters)	Hydroelectricity (000 mwh)	Ethanol (000 kl)
1959	28	—	20	124
1960	25	—	20	187
1961	28	—	7	177
1962	43	—	25	89
1963	31	—	50	116
1964	37	—	100	145
1965	57	—	57	143
1966	69	—	131	149
1967	113	—	109	170
1968	198	—	81	180
1969	206	—	75	185
1970	159	—	75	91
1971	120	5.8	88	66
1972	112	6.9	74	70
1973	138	14.5	62	70
1974	168	19.5	89	71
1975	226	17.2	62	67
1976	235	21.3	53	69
1977	256	17.0	73	77
1978	288	10.6	83	77
1979	288	17.5	104	80
1980	274	17.8	97	84
1981	259	13.3	60	78
1982	541	10.6	43	87
1983	742	8.3	63	86
1984	770	3.4	70	87
1985	868	6.9	54	93
1986	938	5.7	59	97
1987	895	23.9	44	101

Sources: Oil: 1950-58—U.S. Bureau of Mines, *Minerals Yearbook*, various volumes; 1959-67—*Cuba Economic News* 4, no. 34 (1968), p. 4; 1968-87—Cuba, Comité Estatal de Estadísticas, *Anuario Estadístico de Cuba*, various volumes. Natural gas: *Anuario Estadístico de Cuba*, various volumes. Hydroelectricity: United Nations, Statistical Office, *World Energy Supplies 1950-1974* and *Energy Statistics Yearbook*, various volumes. Ethanol: 1959—Cuba Económica y Financiera, *Anuario Azucarero de Cuba 1959*; 1960-61—Cuba, Ministerio del Comercio Exterior, *Anuario Azucarero de Cuba 1961*; 1962—Cuba, Junta Central de Planificación, *Principales Indicadores de la Actividad Económica 1962*; 1963-87—*Anuario Estadístico de Cuba*, various volumes.

year during 1975-80. Dramatic increases in oil production have been recorded in the 1980s, with output reaching 868,000 MT in 1985 and 938,000 MT in 1986. Cuba planned to produce 1 million MT of oil in 1988[2] but fell short of this goal, producing only 717,000 MT.[3] The 1986-90 plan calls for annual oil production to reach 2 million MT by 1990.[4]

Prior to 1968, small quantities of natural gas coproduced with petroleum were generally flared. In that year, commercialization began with the completion of two small gas pipelines connecting the Cristales field with a thermoelectric plant in Ciego de Avila; production statistics were first reported in 1971. Natural gas production peaked at about 20 million cubic meters during 1974-76 and declined steadily thereafter. In 1987, production shot up to 23.9 million cubic meters.

Cuba's hydroelectric resources are limited: Its rivers have low heads, carry relatively small volumes of water, and are subject to uneven rates of flow during the year. Installed hydroelectric generating capacity during the 1950s was approximately 3-4 megawatts (Mw) in six small plants. In the mid-1950s, construction began on a 42.6-Mw hydroelectric plant at the Hanabanilla River; the plant came on-line in 1962-63 and reached full generating capacity in 1967-68.

In the 1950s, before the Hanabanilla plant was completed, annual electricity generated by hydroelectric plants averaged about 14,500 megawatt-hours (Mwh). Their output averaged nearly 100,000 Mwh per year during 1964-67 and declined subsequently to an average of about 70,000 Mwh per year. Output recovered to about 100,000 Mwh in 1979-80. In the 1980s, output of hydroelectric plants has averaged less than 60,000 Mwh per year.

One of the sugar byproducts produced in Cuba in significant quantities is alcohol, and distilleries are commonly integrated with sugar mills. Ethanol production during the 1950s averaged slightly over 100,000 kiloliters (kl) per year, rising to over 150,000 kl per year in the 1960s. Production declined sharply after 1969, to about 75,000 kl per year, as a result of government policies that diverted molasses from ethanol into cattle feed production. Ethanol production averaged about 85,000 kl per year during the first half of the 1980s.

Bagasse, the moist mass of stalks and leaves that remains after sugar cane is ground to extract its juice, is the leading domestic source of energy. Bagasse is used as a fuel exclusively in sugar mills because its bulkiness and low caloric value make it uneconomical to transport.

Official data on bagasse production, and on bagasse used as fuel, are not available. However, we have estimated the availability of bagasse as fuel using official data on sugar cane milled and assuming: (1) a fixed bagasse-to-milled cane ratio of 0.25; and (1) full use of the bagasse as fuel. Following this method, we estimate annual bagasse production at approximately 10.8

million MT per year in 1959-69, 20.0 million MT in the record-setting 1970 sugar harvest, 12.1 million MT in 1971-75 and 15.5 million MT in 1976-80. In 1981-87, production of bagasse has averaged over 17 million MT annually.

Traditionally, fuelwood has been used in Cuban rural homes for cooking and as fuel in some local industries, such as bakeries. In the sugar industry, fuelwood is used to start up sugar mills before bagasse becomes available and to make up for bagasse shortages during the milling process. According to official Cuban statistics, annual fuelwood production averaged 208,000 cubic meters during 1958-59, 1,143,000 for 1960-69, and 1,653,000 for 1970-76. In 1981-87, fuelwood production has averaged more than 2.3 million cubic meters annually.

Charcoal, made from mangrove and other coastal shrubs, was an important home cooking fuel in urban areas until the mid-1940s, when it began to be replaced by kerosene, propane, and electricity. Production of charcoal peaked in 1940 at about 222,200 tons and declined thereafter to 55,000 tons in 1953 and 37,700 in 1958. Production rose again to 55,300 in 1959 and reached a peak between 1960 and 1965, averaging 168,200 tons. Charcoal declined to 99,800 short tons in 1966-69, and 72,500 tons in 1970-76. Data after 1976 are not available.

Table 8.2 estimates the supply of domestic energy during 1959-87 by combining production data converted to a standard unit, thousand metric tons of oil equivalent.[5] For the entire period, bagasse provided nearly 80 percent of domestically produced energy, with its contribution rising as high as 92 percent in 1959 and 88 percent in 1970. Also readily noticeable are the increased importance of oil and the decline in the contribution of ethanol and charcoal to domestic energy supply. In 1981-87, oil contributed about 17 percent of total domestic energy supply, compared with less than 6 percent in the first half of the 1970s and 8 percent in the second half.

Because of the disproportionate importance of bagasse and wide year-to-year fluctuations in bagasse production, no clear trends in the expansion of domestic energy supply can be discerned. During 1959-69, domestic energy supply was basically stagnant, hovering around 2.4 million MT of oil equivalent; excluding 1970, when bagasse output shot up as a result of a record-high sugar crop, domestic supply in the 1970s expanded modestly, at the rate of about 4 percent per annum. In 1981-87, domestic energy supply continued to expand, but underwent severe year-to-year changes; the steady increase in oil production was offset by fluctuations in bagasse availability.

ENERGY IMPORTS

Official Cuban data on imports of energy products for 1959 and 1962-86 are given in Table 8.3. (Data for 1960-61 are not available, as Cuban foreign

Table 8.2. Cuba's domestic energy supply, 1959-1987
(Thousand metric tons of oil or equivalent)

	1959	1960	1961	1962	1963	1964	1965	1966	1967	1968	1969	1970	1971	1972
Oil	28	25	28	43	31	37	57	69	113	198	206	159	120	112
Natural gas	—	—	—	—	—	—	—	—	—	—	—	—	5	6
Hydroelectricity	17	17	6	21	42	84	48	110	91	68	63	63	74	62
Ethanol	65	97	92	46	61	76	74	78	89	94	96	47	35	37
Bagasse	2,000	2,124	2,428	1,642	1,410	1,660	2,267	1,642	2,267	1,892	1,821	3,570	2,303	1,946
Fuelwood	32	71	239	180*	115	143	131	129	133	194	245	198	219	231
Charcoal	35	119	84	109	116	103	105	78	71	62	41	30	41	48
Totals	2,177	2,453	2,877	2,041	1,775	2,103	2,682	2,106	2,764	2,508	2,472	4,067	2,797	2,442

	1973	1974	1975	1976	1977	1978	1979	1980	1981	1982	1983	1984	1985	1986	1987
Oil	138	168	226	235	256	288	288	274	259	541	742	770	868	938	895
Natural gas	12	16	14	18	14	9	15	15	11	9	7	3	6	5	20
Hydroelectricity	52	74	52	44	61	69	87	81	50	36	52	59	45	50	37
Ethanol	35*	35*	35	36	40	40	42	44	41	45	45	45	49	49	51
Bagasse	2,124	2,213	2,267	2,321	2,500	3,000	3,260	2,750	2,960	3,280	3,067	3,498	2,980	3,046	2,987
Fuelwood	249	231	230	278	318*	313	338	330	334	340	325	331	317	322	309
Charcoal	48	50	55	48	48*	48*	48*	48*	48*	48*	48*	48*	48*	48*	48*
Totals	2,658	2,787	2,879	2,980	3,237	3,767	4,078	3,542	3,703	4,299	4,286	4,754	4,313	4,458	4,347

Sources: Oil, natural gas, hydroelectricity, ethanol, see Table 1. Bagasse: Estimated based on data on sugar cane milled from Cuba, Junta Central de Planificación, *Anuario Estadístico de Cuba,* various volumes, and Comité Estatal de Estadísticas, *Anuario Estadístico de Cuba,* various volumes. Fuelwood and charcoal: *Anuario Estadístico de Cuba,* various volumes.
*Estimated.

Table 8.3. Cuban energy imports, 1959 and 1962-1986
(Thousand metric tons)

	1959	1962	1963	1964	1965	1966	1967	1968	1969	1970	1971	1972	1973
Oil and oil products	3,152	4,483	4,088	4,598	4,597	5,058	5,107	5,226	5,675	6,030	6,834	6,691	7,161
Crude oil	1,710	3,720	3,709	3,496	3,483	3,826	3,713	3,851	4,156	4,261	4,757	4,749	5,243
Oil products	1,442	763	379	1,102	1,114	1,232	1,394	1,375	1,519	1,769	2,077	1,942	1,918
Fuel oil	1,199	424	159	766	791	850	975	1,006	975	1,148	1,409	1,315	489
Diesel fuel	17	106	11	186	181	251	312	272	447	483	488	545	1,303
Gasoline	55	176	155	85	74	62	38	17	10	46	86	13	49
Lubricants	171	57	54	65	68	69	69	80	87	92	94	69	77
Coke, coal, and briquets	118	75	97	122	128	102	138	106	121	123	126	140	104
Coal	58	47	69	90	91	64	72	58	69	65	51	52	45
Coke	60	28	28	32	37	38	46	48	52	58	75	88	59

	1974	1975	1976	1977	1978	1979	1980	1981	1982	1983	1984	1985	1986
Oil and oil products	7,786	7,768	8,240	9,245	9,611	9,564	10,203	10,753	11,386	12,107	12,240	13,270	12,891
Crude oil	5,875	5,797	5,783	6,201	6,359	6,131	6,025	6,355	6,247	6,861	7,235	8,046	7,366
Oil products	1,911	1,971	2,457	3,044	3,252	3,433	4,178	4,398	5,139	5,246	5,005	5,224	5,525
Fuel oil	492	1,328	1,487	1,799	2,199	2,287	2,820	2,954	3,438	3,784	3,435	3,421	3,822
Diesel fuel	1,252	481	802	997	751	848	952	1,135	1,298	1,091	1,231	1,311	1,227
Gasoline	49	44	70	151	188	207	250	203	263	249	224	384	391
Lubricants	118	118	98	97	114	91	156	106	140	122	115	108	85
Coke, coal, and briquets	88	98	87	89	140	99	136	134	168	131	140	154	145
Coal	46	57	42	42	60	53	83	75	108	75	85	90	83
Coke	42	41	45	47	49	46	53	59	60	56	55	64	62

Sources: 1959, 1962: Calculated from Junta Central de Planificación, Dirección de Estadística, *Comercio Exterior de Cuba*, volumes for 1959 and 1962. 1963-1986: Junta Central de Planificación, *Boletín Estadístico de Cuba*, various volumes; and Comité Estatal de Estadísticas, *Anuario Estadístico de Cuba*, various volumes.

trade statistics for those two years have not been published. For 1987, data on imports of oil products are not yet available.)

Cuban imports of oil and oil products rose by 78 percent between 1959 and 1969 (from 3.2 million to 5.7 million MT), averaging about 4.5 million MT per year. From 1969 to 1979, imports rose by nearly 69 percent, to 9.6 million MT.

In the 1980s, imports of oil and oil products continued to increase, peaking at 13.3 million MT in 1985 and declining in 1986 to 12.9 million MT. A portion of the liquid fuels imported by Cuba—perhaps as much as 3 million tons in some years—was reexported to Western countries to obtain hard currencies. Imports of coal, coke, and briquets were essentially stagnant during the entire period 1959-80, averaging around 100,000 MT per year. In the 1980s, imports of these products averaged about 140,000 MT annually.

In Table 4, we have attempted to compare the contribution of imports to total apparent supply of liquid fuels (crude petroleum and petroleum products). Since the Soviet Union has been, for all practical purposes, Cuba's exclusive supplier of liquid fuels since mid-1960, we have estimated Cuban imports of oil and oil products in 1960 and 1961 on the basis of Soviet export data.[5]

In the early 1970s, Cuba began to sell oil products, refined in Cuba from imported Soviet crude, in Western Europe for hard currency.[7] Initially, only small volumes of refined products, such as naphtha, were involved. By the 1980s, the volume of reexports had grown to such an extent that during the period 1983-85, reexports of oil products were Cuba's most important hard-currency export, accounting for over 40 percent of such earnings. Over this three-year period, sugar exports—traditionally the main source of export revenue—contributed 21 percent of hard currency earnings and all other exports about 39 percent.

Except for naphtha exports, Cuba does not publish statistics on the volume of liquid fuel reexports. However, using as a very rough approximation the average export price of Cuban shipments to Western Europe, estimates of the volume of non-naphtha oil exports can be made. Those estimates, given in Table 4, suggest that Cuba exported about 720,000 MT of oil and oil products in 1980-81, nearly 1.4 million MT in 1982, 2.9 million MT in 1983, over 2.7 million MT in 1984, nearly 3.4 million MT in 1985, and 2.4 million MT in 1986. These estimates are in line with a reference in the Cuban literature that in 1984, about 23 percent of total oil imports were reexported.[8] (Since 1984 oil and oil product imports were 12.24 million tons [table 8.3], a 23 percent share of reexports translates into an export volume of 2.8 million tons).

Imported products have an overwhelming importance in the apparent

Table 8.4.   Cuba's apparent supply of oil and oil products, 1959-1986
(Thousand metric tons)

	Domestic production	Imports	Exports	Apparent supply	Domestic production as percentage of supply
1959	28	3152	—	3180	0.9%
1960	25	4000	—	4025	0.6
1961	28	4000	—	4028	0.7
1962	43	4483	—	4526	1.0
1963	31	4088	—	4119	0.8
1964	37	4598	—	4635	0.8
1965	57	4597	—	4654	1.2
1966	69	5058	—	5127	1.3
1967	113	5107	—	5220	2.2
1968	198	5226	—	5454	3.6
1969	206	5675	—	5881	3.5
1970	159	6030	—	6189	2.6
1971	120	6834	—	6954	1.7
1972	112	6691	9	6794	1.6
1973	138	7161	25	7274	1.9
1974	168	7786	75	7879	2.1
1975	226	7768	36	7958	2.8
1976	235	8240	49	8426	2.8
1977	256	9245	268	9233	2.8
1978	288	9611	274	9625	3.0
1979	288	9564	241	9672	3.0
1980	273	10203	722	9754	3.0
1981	259	10753	719	10293	2.5
1982	541	11386	1383	10544	5.1
1983	742	12107	2921	9928	7.5
1984	770	12240	2754	10265	7.5
1985	868	13270	3388	10750	8.1
1986	938	12891	2432	11397	8.2

Sources: Domestic production—table 1. Imports—table 3. Exports—Jorge F. Pérez-López, "Cuban Oil Reexports: Significance and Prospects," Energy Journal 8, no. 1 (1987): p. 4, (updated by the author).

supply of liquid fuels (domestic production plus imports less exports). For the first half of the 1960s, domestic production of crude oil accounted for less than 1 percent of the apparent supply of liquid fuels (table 8.4). In the second half of the 1960s, the domestic share rose to 2.4 percent, peaking at about 3.6 percent in 1968-69. After falling below 2 percent in 1971-73, the domestic share rose to nearly 3 percent in 1975-80. In 1982, domestically

Table 8.5.   Cuban energy supply, 1959-1986
              (Thousand metric tons of oil or equivalent)

	Domestic production	Net imports	Apparent supply	Domestic production as percentage of supply
1959	2,177	3,232	5,409	40.2%
1960	2,453	4,070	6,523	37.6
1961	2,877	4,070	6,947	41.4
1962	2,041	4,535	6,576	31.0
1963	1,775	4,156	5,931	29.9
1964	2,103	4,684	6,787	31.0
1965	2,682	4,687	7,369	36.4
1966	2,106	5,129	7,235	29.1
1967	2,764	5,203	7,968	34.7
1968	2,508	5,340	7,848	32.0
1969	2,472	5,759	8,231	30.0
1970	4,067	6,116	10,183	39.9
1971	2,797	6,922	9,719	28.8
1972	2,442	6,780	9,222	26.5
1973	2,658	7,209	9,867	26.9
1974	2,787	7,773	10,560	26.4
1975	2,879	7,831	10,680	27.0
1976	2,980	8,252	11,232	26.5
1977	3,237	9,039	12,276	26.4
1978	3,767	9,435	13,202	28.5
1979	4,078	9,392	13,470	30.3
1980	3,542	9,576	13,118	27.0
1981	3,703	10,128	13,831	26.8
1982	4,299	10,121	14,420	29.8
1983	4,286	9,278	13,564	31.6
1984	4,754	9,593	14,347	33.1
1985	4,313	9,990	14,303	30.2
1986	4,458	10,561	15,019	29.7

Sources: Domestic production—table 8.2; imports—table 3.

produced oil accounted for 5.1 percent of apparent supply, of liquid fuels; during 1983-84, the share of apparent supply accounted for by domestic production was 7.5 percent and slightly over 8.0 percent in 1985-86.

Table 8.5 combines data on domestic energy production and imports to estimate that Cuban apparent energy supply expanded by 178 percent between 1959 and 1986, reaching a level of approximately 15 million MT of oil equivalent in 1986. The expansion pattern was not steady, however, showing both rapid growth spurts and periods of supply stagnation or even decline.

Examination of the data in table 8.5 suggests that during the revolutionary period, the dependence of the Cuban economy on imported energy increased markedly. While domestic energy production accounted for almost 40 percent of total supply in 1959-61, it represented less than 32 percent in 1962-69 and further deteriorated to an average of about 27 percent in 1971-81. (In these calculations, 1970 has been excluded since it was an abnormal year in terms of sugar output and, therefore, of bagasse production.) In what may be a significant positive development for the Cuban economy, domestic energy production during 1982-86 exceeded 30 percent of apparent supply.

ENERGY CONSUMPTION

In comparison with production and import statistics, Cuba's energy consumption statistics are scarcer and weaker. Time series data on total energy consumption, on consumption by sectors, or on consumption by industry are not available. The only official data published regularly refer to the contribution of different energy sources to energy consumption.

Taken together, biomass, ethanol, and oil products fulfill the bulk of Cuba's energy needs. Although there have been small year-to-year fluctuations, their combined share has remained remarkably constant at over 90 percent. The predominance of these energy sources is confirmed by the following data on the contribution of different sources to energy consumption published some time ago in a popular periodical:[9] crude oil, 38 percent; oil products, 27 percent; bagasse, 23 percent; others, 12 percent.

Data on energy consumption by economic sectors are very scanty. The same popular periodical mentioned above reports the following percentage composition of energy consumption:[10] industry, 59 percent; construction, 4 percent; transportation, 15 percent; agriculture, 7 percent; community and personal services, 23 percent; others 4 percent.

Within the industrial sector, the sugar and electricity industries are the largest consumers of primary energy. The sugar industry uses essentially all bagasse production as well as significant amounts of fuelwood, oil products (fuel oil), and electricity.

Between 1958 and 1985, Cuban electric generating capacity expanded over eightfold, from 297 Mw to 2,608.8 Mw.[11] With the exception of the Hanabanilla hydroelectric plant (with generating capacity of 45 Mw, or 1.7 percent of total electric generating capacity) all Cuban electricity generating plants were thermoelectric, fueled primarily with oil products. (Plants in some sugar mills may be fueled with bagasse during the milling period.)

It has been estimated that in the 1970s, electricity generation used from one-fourth to one-fifth of the total oil and oil products consumed.[12]

Relying on official statistics on gross electricity generation and oil consumption per kwh generated, it can be estimated that in 1984, electricity generation consumed nearly 3 million MT, or 30 percent of total consumption, of oil and oil products.[13] Over time, there has been a steady increase in the share of electricity consumed by the industrial sector, which amounted to 44 percent of total electricity produced in 1985, with residential consumption accounting for about 31 percent and commercial activities for about 21 percent.

NUCLEAR POWER

By the end of the century, Cuba plans to generate more than a quarter of its energy from nuclear power,[14] a commitment that could have a significant impact on the energy balance beginning in the 1990s.

As conceived in the early 1970s, the Cuban nuclear power program foresaw construction of the first nuclear power plant on the shore of the Arimao River, several miles northeast of the city of Cienfuegos; construction was to begin in 1977-78 and the plant would be operational by 1985.[15] Hardware and technical assistance, as well as financial assistance, would be provided by the Soviet Union and other member-nations of the Council for Mutual Economic Assistance (CMEA). In early 1978, a new location for the plant was announced, presumably because geological problems were uncovered at the first site. The new location is approximately ten miles southwest of Cienfuegos, in the vicinity of the town of Juragua. A second plant has been slated for the northern part of Holguin providence in eastern Cuba, but there is no information on its exact location. Holguin and Cienfuegos are areas in which extensive industrial investments are planned for the next few years and electricity demand is expected to rise significantly. The province of Holguin, in particular, is the setting for the Cuban nickel industry (Nicaro, Moa, Punta Gorda), a heavy user of electricity. A third plant is to be built in the western part of the island, at a location not yet disclosed.

The reactors for the Juragua nuclear power plant are standard Soviet export 440-MW pressurized water reactors. The prototype was built in Novovoronezh, and the model has been exported by the Soviet Union to Eastern European CMEA countries and to Finland. Information on the type or size of reactors envisaged for the Holguin plant has not been published but, given the total generating capacity of the Cuban electrical grid, it is unlikely that a reactor larger than 440 MW would be feasible. As are all pressurized water reactors, those to be built in Cuba will be fueled with slightly enriched uranium (3 to 4 percent U-235), to be provided by the Soviet Union.

Available information indicates that infrastructure and support facilities for the Juragua plant—including roads, port facilities, housing for prospective employees and Soviet technicians, and a secondary school—have either been completed or are close to completion. Construction of the structure to house the first reactor and associated hardware began in 1983, and the second reactor in 1985.[16] In February 1986, it was reported that the plant's first reactor will become operational in 1990,[17] with the subsequent reactors at the site—up to four—to be completed at two-year intervals.[18]

Considering that over 90 percent of Cuba's oil and oil products needs are met through imports, there is significant incentive for Cuba to diversify its energy sources and to reduce dependence on oil products for electricity generation. The incentive to reduce petroleum consumption has become more acute in the 1980s, as Cuba has worked out an arrangement with the Soviet Union whereby it can reexport, for hard currency, Soviet oil and oil products it has imported but not consumed. President Castro has stated that each 440-MW nuclear reactor will displace 600,000 MT of oil products per year,[19] about one-fifth of the estimated 3 million MT of oil products consumed each year in electricity generation. A Cuban source has reported that the annual "savings" of 2.4 million MT of fuel oil associated with the Juragua plant (four reactors) are equivalent to 70 percent of fuel oil consumption in electricity generation, implying a consumption of 3.4 million MT of fuel oil per year in electricity generation.[20]

ENERGY AND SECURITY

Cuba's heavy dependence on imported energy products is hardly a surprising finding and by itself is devoid of much significance. The real question is whether or not Cuba is vulnerable to disruptions in the flow of energy products. Cuba's reliance on the Soviet Union for virtually all of its energy imports affects Cuba's security and ability to conduct an independent foreign policy. To what extent is Cuba vulnerable to disruptions in energy supplies?

Cuban energy suppliers. Prior to the revolutionary takeover of 1959, Cuban imports of oil products originated exclusively from the West. The United States was the main supplier of crude oil, diesel fuel, gasoline, and lubricants, and along with Aruba and Curacao, it provided the bulk of the fuel oil imported. The United States and Western Europe provided essentially all imported coal and coke.

These import patterns were upset in 1960. On February 13, 1960, the Cuban revolutionary government and the Soviet Union concluded a commercial and payments agreement that provided for the barter of Soviet goods, including oil and oil products, for Cuban sugar and other goods. As oil shipments from the Soviet Union began to arrive in April 1960, international oil companies operating refineries in Cuba (Esso, Texaco, and Royal Dutch Shell) continued their normal purchases of crude from their affiliates—the makings of an oil glut. On May 17, 1960, the Cuban National Bank informed the foreign oil companies that each would have to purchase and process 300,000 MT of Soviet crude during 1960. This move was ostensibly taken to help Cuba's dwindling dollar reserves, since the Soviet crude was obtained through barter and did not require payment in convertible currency. The oil companies balked and the government retaliated on June 29, 1960 by seizing the refineries. From then on, Cuba-U.S. economic and political relations deteriorated rapidly, with the Soviet Union and its allies taking over in supplying Cuba with oil, oil products, and coal.

For the period 1962-86 (as noted earlier, imports data for 1960-61 have not been published; nor are origin-of-imports data available for 1965-66), the Soviet Union provided well over 99 percent of Cuba's imports of liquid fuels. (In 1978, the Soviet Union's share was 96.8 percent; unfortunately, Cuba's import statistics do not specify the source of the non-Soviet imports in 1978.)

A portion of the imports reported as originating from the Soviet Union in fact may have been imported from Venezuela under a swap agreement, signed in November 1976, that resulted in savings on transportation costs. Under the agreement, Venezuela ships up to 20,000 barrels per day (bpd) of crude to Cuba to replace Soviet crude; in return the Soviet Union supplies a matching amount of crude to Spain, Venezuela's traditional customer.[21] In 1977, the swap amount was limited to 5,000 bpd; it rose to 10,000 bpd in 1978-80 and reached the full 20,000 bpd by 1984.[22] Apparently the Soviet Union requested an increase in the volume of petroleum in the swap but Venezuela did not accede.[23] However, in mid-1985 it was reported that Venezuela had agreed to double the volume of the swap with the Soviet Union to 40,000 bpd.[24] Reports of a similar (triangular) sales agreement involving Mexico, Cuba, and the Soviet Union have circulated in the oil trade press since 1978, but the existence of such a deal has not been confirmed.[25]

Although Cuban statistics are less clear on this point, it appears that since the mid-1960s, the Soviet Union has been the source of most Cuban coke and coal imports. During 1962-64, the only years for which coke and coal import statistics by country of origin are available, the Soviet Union provided less than 40 percent by weight of total imports; Albania and North Vietnam supplied anthracite, and Poland bituminous coal and coke. Beginning in 1967, the volume of Soviet exports of these products to Cuba (as reported in Soviet foreign trade handbooks) coincides very closely with Cuban imports.[26] Official Polish trade data last record bituminous coal shipments to Cuba in 1965 and coke shipments in 1967[27]

Adoption of nuclear power for electricity generation will reduce the importance of liquid fuels in the energy balance. However, since Cuba has no known deposits of fissionable materials and no facilities to enrich uranium, the country will have to rely on imported enriched uranium from the Soviet Union. In other words, while adoption of nuclear power for electricity generation will broaden the energy base, the vulnerability to external supply manipulation will prevail.

Energy and foreign policy. There is evidence that in 1967-68 the Soviet Union used the "oil weapon" to force changes in Cuban foreign policy.[28] Beginning in 1965, differences between Cuba and the Soviet Union regarding the correct revolutionary strategy vis-á-vis Latin America intensified. While Cuba favored armed struggle and support of guerrilla movements to gain influence, the Soviet Union was the proponent of a more traditional strategy of normalizing diplomatic and commercial relations with Latin American governments. At the first conference of the Latin American Solidarity Organization (OLAS), hosted by Cuba in July-August 1967, Castro attacked Soviet support of established governments and challenged Soviet revolutionary commitments.[29]

Meanwhile, negotiations between Cuba and the Soviet Union on a trade protocol for 1968, which should have been concluded in the fall of 1967, dragged on. Reportedly, Soviet oil shipments in the fall and winter of 1967 became irregular and began to have an adverse effect on the economy. On January 2, 1968, Castro spoke publicly about the fuel shortages and imposed a strict rationing system. In his speech, Castro noted that the Soviet Union had made a "considerable effort" to provide Cuba with fuel but "by all indications the current possibilities of that nation to meet Cuba's growing needs are limited."[30] At the same time that oil deliveries to Cuba slowed down, the Soviet press announced that production of oil and oil products "was so high that the Soviet Union would increase exports of these products to Latin America, including Brazil and Chile, two nations with which Cuba had bitter disputes."[31]

Through the spring of 1968, the two sides hardened their positions, with the Soviet Union replacing its ambassador to Cuba and Castro denouncing a pro-Soviet "microfaction" of members of the Cuban Communist party and ordering their expulsion. Energy conservation and the need to do more with fewer resources to save the revolution became the rallying point of the Cuban leadership. The Cuban-Soviet trade protocol, finally signed in late March 1968, provided for a modest increase in trade but no immediate relief to the oil shortages. On August 23, 1968, however, Castro supported publicly the Soviet occupation of Czechoslovakia and entered into a new accommodation with the Soviets. From here on, Cuban foreign and domestic policies have been in remarkable agreement with those of the Soviet Union, and Soviet fuel exports to Cuba have flowed uninterruptedly.

Nuclear nonproliferation. An argument could be made that the only major foreign policy issue on which Cuban and Soviet views do not mirror each other is nuclear nonproliferation. Cuba has adamantly refused to adhere to either the Treaty of Tlatelolco or the Non-Proliferation Treaty. However, Cuba is a member of the International Atomic Energy Agency (IAEA) and has signed a safeguards agreement with the IAEA covering the Juragua plant.

Shortly after the Cuban missile crisis of October 1962, the heads of state of five Latin American countries (Bolivia, Brazil, Chile, Ecuador, and Mexico) called for hemispheric consultations to create a nuclear weapons–free zone in Latin America. After prolonged negotiations, such an agreement went into effect in 1969, when the requisite number of nations had ratified it. The agreement has not reached hemispheric coverage, however, since some eligible nations (Cuba, Guyana, St. Lucia, Dominica, and Belize) have not yet signed and one nation (Argentina) has not ratified. All of the nuclear weapons states have signed protocol II, binding them not to deploy nuclear weapons in Latin America.

In August 1965, the Cuban government stated that it would not participate in the negotiations of an agreement to denuclearize Latin America because the United States deploys nuclear weapons and maintains military bases in the area. This position was further explained in a formal response to the president of the Preparatory Commission for the Denuclearization of Latin America, in which Cuba set forth three conditions for participation in the negotiation of such an agreement. Cuba demanded that the United States: (1) remove its military base in Cuban territory at Guantanamo; (2) dismantle military bases in Latin America and stop deploying nuclear weapons in Puerto Rico, the Virgin Islands, and the Panama Canal; and (3) discontinue its aggressive policies toward Cuba. Despite numerous efforts by the secretary general of the Agency for the Prohibition of Nuclear Weapons in Latin America (OPANAL) to engage Cuba in meaningful discussion, the latter has steadfastly maintained these conditions.

In 1978, the Soviet Union reversed a decade of opposition to the

Tlatelolco treaty and announced that it would sign protocol II (open to signature by nuclear weapons states). This decision was significant since the Soviet refusal to sign the agreement had appeared to be motivated by solidarity with the Cuban position. There was speculation that as a result of the close relationship between former Mexican President Lopez Portillo and Castro (Mexico is the prime force behind the Latin American nuclear-free zone movement), Cuba might be persuaded to adhere to the Tlatelolco treaty, but so far its position is unchanged.

The Non-Proliferation Treaty. In early 1968, the U.N. General Assembly opened discussion on a multilateral agreement to curb the spread of nuclear weapons. According to the instrument that resulted—the Treaty on the Non-Proliferation of Nuclear Weapons (NPT)—have-not states agreed not to develop nuclear weapons in return for a commitment from nuclear weapons states to transfer, without discrimination, nuclear technologies for peaceful purposes. All states without nuclear weapons agreed to place their nuclear installations under a system of safeguards administered by the IAEA and to permit IAEA to inspect nuclear facilities.

During the U.N. debate of the NPT, Cuban Foreign Minister Raul Roa stated that "Cuba would never give up its inalienable right to defend itself using weapons of any kind, despite any international agreement."[32] This view, which reflected in part Cuba's hostility toward the United States, was consistent with the earlier Cuban rejection of the Limited Test Ban Treaty of 1963 and the Outer Space Treaty of 1967 and was also a slap at the Soviets, who, with the United States, were the cosponsors of the NPT. Cuba's rejection of the NPT, which is grounded on the same reasons as the Tlatelolco treaty, remains unchanged; Cuba's views were restated by Vice President Carlos Rafael Rodriguez at the May 1978 U.N. General Assembly Special Session on Disarmament.[33]

More recently, Fidel Castro Diaz-Balart, executive secretary of the Cuban Atomic Energy Commission, summarized Cuba's rationale for not signing the NPT or the Treaty of Tlatelolco: "As is well known, Cuba has abstained from entering into these international agreements—without questioning their importance—because of issues of principle: the permanent hostility, threats, blockade, and aggressions suffered during the last twenty-six years from the only nuclear-weapon state in the hemisphere, the United States, who also illegally usurps, against our will, a part of our territory. It is clear that as long as these circumstances remain, it would not be either dignified or acceptable for our nation to make unilateral concessions."[34]

The chief objective of the IAEA international safeguards system is to ensure that nuclear equipment and fissionable materials earmarked for peaceful purposes are not diverted for military purposes. This end is accomplished through an early detection system that relies heavily on review of records maintained by nuclear reactor operators and periodic on-site verification of records by IAEA officials. Signatories to the NPT must place all their nuclear facilities under IAEA safeguards; nonsignatories may voluntarily enter into safeguards agreements with the IAEA as well.

On May 5, 1980, Cuba and the IAEA entered into an agreement to safeguard the Soviet nuclear plant to be built in Juragua and its fissionable materials. According to Section 2 of the agreement, "the Government of Cuba undertakes that none of the [hardware and fissionable materials transferred for the power plant] . . . shall be used for the manufacture of any nuclear weapon or to further any other military purpose or for the manufacture of any other nuclear explosive device."[35] Like all standard safeguards agreements, the Cuba-IAEA pact also provides for an inventory of nuclear hardware and fissionable materials and their movements into or out of Cuban territory, with inspections by IAEA officials.

Traditionally, the Soviet Union has been a strong supporter and practitioner of nuclear nonproliferation; it was a moving force behind the NPT, has openly encouraged its adoption by other nations, and has signed Protocol II of the Treaty of Tlatelolco. As a member of the Nuclear Suppliers' Club, or London Club, the Soviet Union has entered into a cartel with fourteen Western and Eastern nations to regulate the sale of nuclear hardware and fissionable materials. In exports to CMEA countries and to Finland, the Soviet Union maintains control over fissionable materials and states that it intends to retrieve the irradiated fuel for reprocessing. This arrangement is described by a Soviet official as follows: "Nuclear power development in the CMEA member countries . . . . [has] led to an arrangement whereby the U.S.S.R. carries out isotopic enrichment in uranium-235, fabricates and supplies to CMEA countries "fresh" fuel for their nuclear power plants and takes back the spent fuel for reprocessing. All this creates favorable conditions for compliance with the provisions and requirements of the NPT."[36] Further, Soviet export reactors are the type from which it is extremely difficult to divert fissionable fuel.[37]

In effect, the sale of nuclear technology and materials to Cuba represents an important departure in Soviet nuclear export policy. The Soviet Union previously has insisted that recipients of nuclear hardware and fissionable materials sign and ratify the NPT. Finland and all the Eastern European CMEA countries are signatories, as are Libya and Iraq, two countries often mentioned in connection with Soviet exports of nuclear hardware. It is unclear how heavily the Soviet Union has pressured Cuba to sign the NPT. On the one hand, given Cuba's strong, overt opposition to the NPT, coercion to sign the agreement would harm Cuba's international image, providing further evidence that Cuba does not have a foreign policy independent of Moscow. On the other hand, Cuba's rejection of the NPT does not fit with the Soviet Union's nonproliferation record. The Soviet Union appears to have opted for a compromise: permit Cuba to remain outside the NPT but make sales contingent on Cuba's conclusion of IAEA safeguards agreements, and extend to Cuba the policy whereby the irradiated fuel is eventually returned to the Soviet Union for reprocessing.

Nuclear safety. A final security-related concern is the possibility of nuclear accidents. The safety issue is important since an accident at the nuclear power plant being built at Juragua could affect not only the Cuban territory but other neighboring states as well.

The April 1986 accident at the Chernobyl nuclear power plant in the Soviet Union dramatically illustrated the level of destruction possible at a commercial nuclear power facility. Chernobyl also highlighted the differences in approach to nuclear safety in the West and in the Soviet Union. In the West, the cornerstone of nuclear reactor safety is to build redundant systems to counteract a wide range of possibilities. The Soviet approach is to take the greatest care in the design and construction of plant and equipment ("engineered safeguards") and to minimize redundant systems by limiting them to "credible" (all but low-probability) events.

It is clear that the Number 4 reactor at Chernobyl, the unit involved in the April 1986 accident, lacked the pressure-tight containment structure common for nuclear power plants in the West. Had this reactor been so equipped, the containment structure might have confined much—perhaps most—of the radioactive materials that were spewed into the atmosphere. During the most serious nuclear accident in the United States, the 1979 malfunction at Three Mile Island, a prestressed concrete containment structure played a crucial role in preventing the escape of radioactive material, although the enormous amount of energy released at Chernobyl might well have breached even the strongest containment structure.

There is no definitive information on whether the nuclear plants to be built in Cuba will be equipped with three common safety features in Western-built plants: containment structures, emergency cooling systems, and up-to-date instrumentation. Until a few years ago, Soviet-supplied nuclear power plants lacked the secondary containment shells routinely built around Western-supplied plants. In accord with the Soviet view of the impossibility of loss-of-coolant accidents and core meltdowns, Soviet reactors also lacked redundant systems favored in the West for cooling down reactors in case of accidental overheating. Finally, there have been reports that the control rooms of Soviet power plants are equipped with instruments and computers obsolete by Western standards.

The No. 5 reactor at Novovoronezh in the Soviet Union, a 1,000-MW pressurized water reactor that began operations in early 1981, was constructed with a containment structure. Some American observers suggested that the containment unit at the Novovoronezh site was related to the large size of the reactor and was not necessarily indicative of a shift in Soviet philosophy. Other observers suggested that the motivation behind the containment structure was the Soviet intention to gain a larger share of the nuclear export market outside CMEA. Indeed, a high-ranking Soviet nuclear safety expert is reported to have said in 1978 that "The fifth unit at Novovoronezh will have a containment structure as an experiment. But it is a vain expenditure of money."[38] However, there is evidence that at least some of the Soviet VVER-440 reactors currently under construction in Eastern Europe will be equipped with some sort of containment structure, although the exact specifications are not known.[39]

Scattered information suggests that, like these VVER-440s, the Juragua reactors in Cuba will be built with containment structures. References to safety features of the plant by the Cuban government officials are vague,[40] and do not permit a conclusive determination as to whether the containment will meet Western standards. It is clear, however, that some sort of massive structure is being built around the reactors. If, as suggested, the Cuban reactors will have containment, it is puzzling that the Cuban government has not chosen to make that clearly known to its citizens and those in neighboring states. Nor is there solid information available on whether emergency cooling systems or modern instrumentation will be part of the Juragua plant.

The destruction from a nuclear accident of Juragua would depend, of course on the severity of the accident, the extent of the radioactive leakage, and wind direction and speed. Most likely to be affected is the city of Cienfuegos (population over 100,000), approximately ten to fifteen miles from the site. Beyond that possibility, the Juragua plant lies within 150 miles of twenty-nine of the forty-seven urban centers with more than 20,000 inhabitants in 1985, including Havana as well as provincial capitals Santa Clara, Matanzas and Camaguey.[41] In a major accident, the U.S. mainland[42] and other Caribbean nations could also be at risk.

The Chernobyl accident raised U.S. consciousness of the implications of a nuclear accident in Cuba. Immediately after the Soviet Union officially announced that an accident had taken place at Chernobyl, members of the U.S. Congress from Florida tried to learn more about the safety of the Juragua plant. The approaches ranged from letters to the Organization of American States to seek its involvement on behalf of the highest safety standards, to appeals to President Reagan and Soviet General Secretary Gorbachev for immediate halt to the construction at Juragua, to requests to Cuban President Fidel Castro for assurances about the safety of the plant. President Castro chose in June 1986 to respond directly to one U.S. congressman discussing the features of the plant.[43] Later, at a national address on July 26, 1986, President Castro spoke about the safety of the nuclear reactors being built in Cuba: "[The Juragua plant] is of Soviet technology, built with painstaking quality and with the highest safety indexes, so that we can affirm that that nuclear power plant will be safer than any nuclear power plant built in the United States, and with a proven technology at the international level."[44]

Since July 1986, the United States and the Soviet Union have been holding discussions on the safety of the Juragua plant.[45] After a meeting in September 1986, U.S. Secretary of Energy Herrington stated that Soviet authorities had agreed to supply the United states with technical data on the reactors being built at Juragua.[46] There have been no public reports on the status of the information exchange agreement.

ENERGY AND THE ECONOMY

In September 1974, at a time when Western developed and oil-importing developing countries alike were struggling with the adjustment problems caused by a four-fold increase in the world price of oil, Castro boasted that Cuba "with the generous help of the Soviet Union, has not known the energy crisis."[47] Castro was correct in his assertion; however, in retrospect, it would have been more accurate to say that Cuba had *not yet* experienced the energy crisis. It could be argued that because of subsidized Soviet oil, the Cuban economy has been slow to adjust to the realities of high-priced energy. Uncertainties about the willingness and/or ability of the Soviet Union to continue to supply fuel at concessional terms over the long run present a serious challenge for the Cuban economy.

Pricing in socialist foreign trade. Socialist countries that are members of the CMEA follow different pricing practices in trading with Western nations and with other socialist nations. In dealing with Western countries, as a rule they accept and adjust to world market prices. In intra-CMEA trade, however, external prices are generally based on average world market prices for an extended earlier time period, typically five years. These average prices may be adjusted by mutual consent to account for such factors as quality differences and differentials in transportation costs. The official rationale for the use of such averages to determine intra-CMEA prices is the desire to delete from the pricing system the capitalist influences of speculation, business cycles, and monopoly.

Intra-CMEA oil pricing. When OPEC nations increased oil prices in 1973-74, Cuba and the rest of the CMEA nations did not feel the impact at once since they received the bulk of their supplies from the Soviet Union, presumably at prices fixed for the entire period 1971-75. In his first major speech after the 1973 oil embargo, Castro alluded to the rise in oil prices in the world market but did not indicate concern about their impact on the Cuban economy.[48] On December 21, 1973, the Cuban official newspaper *Granma* carried a report from the Soviet news agency Tass in which the Soviet vice-minister for foreign trade was quoted as saying: "The Soviet Union will fulfill completely its commitments with the Socialist countries regarding deliveries of crude oil and oil products and it will not increase prices despite the sharp rise of prices in the world market."[49]

However, the Soviets called a special meeting of the CMEA Executive Committee in January 1975 to review the methodology for pricing raw materials, including oil. The result of the meeting was the adoption, at the Soviet Union's insistence, of a new pricing policy whereby intra-CMEA petroleum prices would be adjusted annually rather than only during the first year of each five-year period, as had been customary. For 1975, the intra-CMEA price was set on the basis of world market prices during the three-year period 1972-74; thereafter, prices would be adjusted to reflect world prices in the previous five-year period. These modifications of intra-CMEA pricing rules, which were made retroactive to January 1, 1975, appear to have caught most CMEA members by surprise. Oil price adjustments had been expected for 1976-80, but not for 1975, the last year of the 1971-75 plan period.

It should be stressed, however, that even with the change in pricing policy, the intra-CMEA petroleum price in 1975 was much lower than the world market price. According to estimates, Cuba paid approximately $5.82 per barrel for Soviet crude and products in 1975, compared with a prevailing world market price for crude of approximately $11.53 per barrel.[50] Similarly, in 1980, the year of the second oil price shock, Cuba paid approximately $15.30 per barrel for Soviet oil—roughly half the world market price of around $30 per barrel.[51] In 1981 and 1982, the Cuban import price for Soviet oil continued to increase (to about $21 per barrel) but remained substantially below the world market price of about $33 per barrel in 1982. After that, however, world market prices plummeted by more than 50 percent, to about $15 per barrel in early 1986. It appears that the price advantage for oil purchases from the Soviet Union disappeared by 1983 and that, at least during 1984-86, Cuba paid a higher price for Soviet oil ($28.50 per barrel in 1985 and $31.90 per barrel in 1986) than the world market price. However, since Cuba obtained oil from the Soviet

Union through barter—without having to use hard currencies—it could benefit from reexporting part of the Soviet oil for hard currency.

The effect of the intra-CMEA moving average pricing formula for oil effective since 1975 is to pass through changes in world market prices, but with a lag. The delayed impact shows up in data on the value of energy imports relative to total imports. Despite the 1973-74 OPEC oil price increases, Cuban imports of energy products (primarily oil and oil products) in 1974 represented a lower percentage of the value of total merchandise imports than in 1973 (9.0 percent versus 11.1 percent), approximating the share they had held in 1963-71. While energy products, share of total imports began to rise in 1975, the increase was quite gradual, except for a significant jump in 1978. Nonetheless, energy imports accounted for nearly 23 percent of the value of total imports by 1981 and for over 33 percent by 1986. In 1986, energy was the single most important category of imports, surpassing capital goods. (It should be recalled, however, that some of these imports—as much as one-fifth—are in fact reexported and have become one of the key sources of hard currency earnings.)

Cuba's imports of oil and oil products, though small relative to Soviet exports to CMEA countries, have been rising steadily. From 1962 through 1965, the Soviet Union provided Cuba with an average of 4.4 million MT of crude and petroleum products per year. From 1966 through 1970, the yearly average was 5.4 million MT per year; from 1971 through 1975, 7.2 million MT; and from 1976 through 1980, 9.5 million MT; and during 1981-86, over 12 million MT. In the 1980s, Cuba took an estimated 6.5 percent of total Soviet petroleum exports and more than 13 percent of Soviet exports to CMEA.[52]

After the mid-1960s, the Soviet Union began to encourage CMEA members to diversify sources of oil supplies and to import from Middle Eastern and African suppliers.[53] In the early 1970s, increased domestic and export demand for Soviet oil, together with a slower rate of production increases, raised questions among some analysts as to whether or not the Soviet Union could satisfy its own and CMEA's future oil demands.[54] A controversial CIA study released in early 1977 predicted that the Soviet Union would become a net oil importer by 1985.[55] While it is generally acknowledged that the CIA forecast was too pessimistic, there is wide agreement among experts that the rate of increase in Soviet production has slowed considerably. The evidence is quite convincing that the Soviets are having to invest ever larger amounts of capital for shrinking marginal returns in oil production.[56]

In any case, there is a question as to whether the Soviet Union will continue to supply petroleum to CMEA in soft currencies. Clearly, the

Soviet Union could obtain much-needed convertible currencies if it sold oil in the world market rather than to its allies.

After taking a hard-line stance at the thirty-fourth CMEA session in June 1980, the Soviet Union in the end agreed to increase energy deliveries to CMEA in 1981-85 by 20 percent over the volume received in the previous five-year period. Apparently the Soviet Union's intention was not to increase petroleum shipments by 20 percent, but rather to increase total energy supply (petroleum, coal, electricity, and natural gas) by this amount while maintaining oil shipments at 1980 levels.[57] In 1982, the Soviet Union reduced soft-currency oil shipments to Eastern Europe (except Poland) by about 10 percent, and redirected exports to hard-currency markets. This move forced Eastern Europe to implement stringent energy conservation programs and to seek other sources of oil. Meanwhile, between 1981 and 1984, Soviet oil exports to hard-currency Western markets grew by about 60 percent.[58]

Not only was Cuba spared the supply reductions experienced by Eastern Europe, but Cuban petroleum imports from the Soviet Union actually increased from 11.7 million to nearly 13.1 million MT between 1982 and 1986, a 12 percent gain.

No official information is available about the intended level of Soviet petroleum shipments to Cuba in the future. However, one source has reported that the Soviet Union has committed to supplying 11 million MT annually, while another has indicated that Cuba will be permitted to export, for hard currency, any amount it does not consume of the 10 million MT per year it is to receive from the Soviet Union.[59]

Cuban apparent consumption of petroleum averaged around 10.5 million MT annually during 1981-86. With domestic production expected to amount to about 1 million MT per annum, imports from the Soviet Union of some 10 million or 11 million MT will more than cover domestic needs and leave a sizable portion for reexport. However, only under the most optimistic domestic production and import scenarios would Cuba be able to approach its 1983-85 exports of about 3 million MT of petroleum. Realistically, foreseeable gains in domestic oil production and conservation would probably only add marginally to exportable oil balances.

Energy conservation. Revolutionary Cuba's overall record in energy conservation has been far from impressive. This is not surprising given the frequent (and contradictory) changes in energy policy, as well as lack of economic incentives to conserve. In the 1980s, the possibility of reexporting imported petroleum not consumed domestically has given a new impetus to energy conservation. Energy conservation has been elevated in terms of national priorities.

The five-year plan for 1986-90 calls for special efforts to reduce by 1990 the energy consumption:national production relationship by 10 percent from that in 1985.[60] Some of the energy conservation measures put into place in the 1980s appear to be having a positive impact. For instance, the Cuban National Bank reported that in 1984, approximately 180,000 MT of petroleum was saved nationwide.[61] In 1985, the energy conservation program reportedly reduced consumption by an additional 220,000 MT[62]

Energy substitution and nuclear power. Given the uncertainties regarding the future supply and price of Soviet oil and oil products, it is clearly in Cuba's interest to diversify its sources of energy and to move away from reliance on oil. However, Cuba's short-term options are limited. Barring spectacular breakthroughs in harnessing solar energy, nuclear energy appears to be the only viable new energy source to which Cuba could turn in the near future. Cuba's leadership has rationalized the decision to turn to nuclear power as follows:

> If one considers the fact that a large quantity of oil, which is almost completely imported, is used to generate electrical power in such a volume, and the prices for this kind of fuel at present and in the future are taken into account, as well as the cost of transportation, it then becomes obvious why the utilization of nuclear electric power has been taken as the basis for electrical power engineering development in our nation as the sole non-traditional technology which has been developed at the present time to a sufficient extent for the needs of the expanding national economy.[63]

Indeed, electricity generation, which uses about 30 percent of total liquid fuels, is the most obvious candidate for sizable reductions in consumption. Beginning in the 1990s, the 440-MW nuclear reactors at Juragua—which reportedly will displace 600,000 MT of oil products annually—could make a significant contribution to the overall energy balance. However, while the Juragua plant is being built, Cuba is also in the process of adding thermoelectric generating capacity: 1130 MW during 1986-90 and at least 450 MW during 1991-95.[64] These additions will keep the electrical industry a major user of oil products through the end of the century.

While nuclear power could contribute positively to the Cuban energy balance, it will not solve Cuba's energy vulnerability. To be sure, if the current construction timetable is adhered to, nuclear reactors could represent as much as one-fourth of electric generating capacity by 1995 and 40 percent by the year 2000.[65] Clearly, nuclear power will reduce Cuba's dependence on oil for electricity generation and thus its vulnerability to an oil import cutoff. However, since fissionable materials for nuclear power

will also be imported—like oil, from the Soviet Union—Cuba's vulnerability to external energy supply interruptions will remain high even if
the energy mix is changed by nuclear power.

NOTES

The opinions expressed in this paper are those of the author and do not necessarily
represent the views of the U.S. Department of Labor.

1. This section draws heavily on Jorge F. Pérez-López, "Energy Production,
Imports and Consumption in Revolutionary Cuba," *Latin American Research Review*
16, no. 3 (1981).

2. "Producciones físicas para 1988," *Granma*, Dec. 31, 1987, p. 4.

3. Banco Nacional de Cuba, *Economic Report*, Havana, May 1989, p. 27.

4. *Lineamientos económicos y sociales para el quinquenio (1986-1990)*, (Havana:
Editora Política, 1986), p. 78.

5. Standard factors from United Nations, *World Energy Supplies 1950-74*, Series
J, no. 19 (New York: United Nations, 1976), and U.S. Federal Energy Administration,
Energy Interrelationships (Washington: U.S. Government Printing Office, 1977), have
been used to convert production of natural gas, ethanol, hydroelectricity, fuelwood,
and charcoal to petroleum equivalents. For bagasse, we have estimated that 5.6 MT of
bagasse (50 percent moisture content) are equivalent in thermal value to 1 MT of crude
petroleum of medium gravity. See Rámon Quesada González, "Consumo de energía
térmica en la producción de azúcar cruda," *Cuba Azúcar* (July-September 1969): p. 6,
and Magaly E. Rodriguez and Raul Gutiérrez, "Estudio sobre la calidad del bagazo,"
ATAC 37, no. 1 (January-February 1978): p. 22.

6. Soviet export data was obtained from Ministerstvo Vneshnei Torgovli SSSR,
Vneshniaia Torgovlia SSSR v 1973g, Moscow, 1974.

7. The discussion of oil reexports draws heavily on Jorge F. Pérez-López, "Cuban
Oil Reexports: Significance and Prospects," *Energy Journal* 8, no. 1 (1987).

8. Serafin Hernández Cruz and Daniel Puentes Alba, "El ahorro de energía:
nueva fuente de recursos energéticos," *Revista Estadística* 9, no. 20 (April 1987): p. 12.

9. Raul Palenzuelos, "¿Ave fénix del desarrollo?" *Bohemia* 74, no. 38 (Sept. 17,
1982): p. 34.

10. *Ibid.*

11. The figure for 1958 is from Fidel Castro, "Discurso en el acto de clausura del
Primer Forum Nacional de Energía," *Bohemia* 76, no. 50 (December 14, 1984): p. 52;
the figure for 1985 is from Comité Estatal de Estadisticas, *Anuario Estadistico de Cuba
1985*, p. 255.

12. The lower consumption figure is from *Granma*, May 3, 1971, p. 3, and the
higher from Lawrence H. Theriot, "Cuba Faces the Economic Realities of the 1980s",
in U.S. Congress, Joint Economic Committee, *East-West Trade: The Prospects to 1985*
(Washington: U.S. Government Printing Office, 1982), p. 119.

13. See Jorge F. Pérez-López, "Nuclear Power in Cuba after Chernobyl," *Journal of
Interamerican Studies and World Affairs* 29, no. 2 (Summer 1987): p. 108.

14. Fidel Castro Díaz-Balart, "La energía nuclear en Cuba: sus perspectivas y las
realidades del mundo de hoy," *Cuba Socialista*, no. 15 (May-June 1985): p. 38.

15. For a more detailed analysis of the nuclear power program, see Pérez-López,
"Nuclear Power in Cuba after Chernobyl"; idem, "Nuclear Power in Cuba: Opportunities and Challenges," *Orbis* 26, no. 2 (Summer 1982).

16. Castro Díaz-Balart, "La energía nuclear en Cuba," p. 46.

17. Joaquin Oramas, "Pondran en marcha a en 1990 el primer reactor VVER-440 en
la Electronuclear," *Granma*, March 21, 1986, p. 1.

18. Joaquín Oramas, "La magia del átomo," *Cuba Internacional* 18, no. 203 (October 1986): p. 20; Jorge Petinaud Martínez, "Nuestra obra del siglo crece en Juraguá," *Nucleus*, no. 0 (1986): p. 25.

19. Castro, "Discurso en el acto de clausura," p. 55.

20. Oramas, "La magia del átomo."

21. "Francas y a veces rudas las conversaciones en el Kremlin," *El Nacional* (Caracas), Nov. 27, 1976, p. D3; see also David Binder, "Venezuela and Soviet Reach an Agreement on Oil," *New York Times*, Dec. 10, 1976, p. D3.

22. George W. Grayson, "Soviet-Venezuelan oil imports deal," *Petroleum Economist* 52, no. 2 (February 1985), p. 60.

23. "Venezuela Pessimistic on Oil Deal," *Journal of Commerce*, February 20, 1980, p. 33.

24. "Venezuela duplicará su petróleo a Cuba roja," *Diario las Americas*, June 13, 1985, p. 1A.

25. See, for example, "Mexico to Supply Oil to Cuba In Swap Deal with Moscow," *Washington Post*, June 4, 1978, p. C10; and Grayson, "Soviet-Venezuelan Oil."

26. Ministerstvo Vneshnei Torgovli SSSR, *Vneshniaia Torgovlia SSR*, various volumes.

27. Glowny Urzad Statystyczny, *Rocznik Statystyczny Handlu Zagranicznego 1965* (Warsaw, 1966) and volumes for 1966, 1967, 1968, 1970, 1973, 1974.

28. It has also been suggested that in the early 1960s, during the Sino-Soviet rift, the Soviets delayed oil shipments to convince Cuba of the inadvisability of trying to follow an independent path between the Soviets and the Chinese. According to Padula, "The Russians broke the oil bridge to bring Castro to his senses. U.S. Navy patrol planes observed Soviet tankers motionless in the high Atlantic. Castro responded angrily, saying that, if necessary, Cuba would abandon the machine age and return to ox-carts in order to preserve its dignity and independence. But after a time, Cuba backed down and sided with the Soviets." See Alfred Padula, "Cuba's Pending Energy Crisis," *Caribbean Review* 8, no. 2 (Spring 1979): p. 6.

29. For treatments of these events, see Edward Gonzalez, "Relationship with the Soviet Union," in Carmelo Mesa-Lago, ed., *Revolutionary Change in Cuba* (Pittsburgh: Univ. of Pittsburgh Press, 1971), pp. 90-97; Maurice Halperin, *The Taming of Fidel Castro* (Berkeley: Univ. of California Press, 1981), chapters 32-39; Carla Anne Robbins, *The Cuban Threat* (Philadelphia: Institute for the Study of Human Issues, 1983), pp. 156-65; and W. Raymond Duncan, *The Soviet Union and Cuba* (New York: Praeger, 1985), pp. 70-79.

30. *Granma*, Jan. 3, 1968, p. 3.

31. Jorge I. Domínguez, *Cuba: Order and Revolution* (Cambridge: Harvard Univ. Press, Belknap Press, 1978), p. 162.

32. *Granma*, May 14, 1968, p. 4.

33. *Granma Weekly Review*, June 11, 1978, pp. 2-3.

34. Castro Díaz-Balart, "La energía nuclear en Cuba," p. 82.

35. "The Text of the Agreement of 5 May 1980 between the Agency and Cuba Relating to the Application of Safeguards in Connection With the Supply of a Nuclear Power Plant," IAEA information circular 281, June 1980, p. 2.

36. A.F. Panasenkov, "Co-operation among CMEA Member Countries in the Development of Nuclear Energy: Its Role in the Implementation of the NPT," *IAEA Bulletin*, August 1980, p. 97.

37. It is generally agreed that extraction of irradiated fuel from pressurized water reactors, such as those exported by the Soviet Union, can be readily detected. See, for example, Joseph L. Nogee, "Soviet Nuclear Proliferation Policy: Dilemmas and Contradictions," *Orbis* 24, no. 4 (Winter 1981), p. 756.

38. Quoted in Wil Lepkowski, "U.S.S.R. Readies Takeoff in Nuclear Power," *Chemical and Engineering News*, Nov. 6, 1978, p. 33.

39. See John Kramer, "Chernobyl and Eastern Europe," *Problems of Communism* 35, no. 6 (November-December 1986): pp. 41,43.

40. For a sampling of Cuban references to safety features of the Juragua plant, see Pérez-López, "Nuclear Power in Cuba after Chernobyl," pp. 91-93.

41. Author's estimates based on Cuba Junta Central de Planificación, Dirección Central de Estadística, *Densidad de Población y Urbanización* (Havana: Editorial Orbe, 1975), pp. 59-60; and Academia de Ciencias de Cuba, *Atlas Nacional de Cuba*, Havana, 1970, pp. 122-23.

42. Statement by Phillip G. Berman, nuclear weapons analyst, U.S. Defense Intelligence Agency, as reported in U.S. Congress, Committee on International Relations, Subcommittee on International Political and Military Affairs, *Soviet Activities in Cuba*; Parts VI and VII (Washington, D.C.: U.S. Government Printing Office, 1976), p. 114.

43. "Cartas intercambiadas por el representante norteamericano Michael Bilirakis y Fidel sobre la electronuclear de Cienfuegos," *Granma*, August 25, 1986, p. 1.

44. Fidel Castro, "Discurso pronunciado el 26 de julio de 1986," *Granma*, July 29, 1986, p. 2.

45. See U.S. Senate, Committee on Governmental Affairs, Subcommittee on Energy, Nuclear Proliferation, and Governmental Processes, *Cuban Nuclear Reactors* (Washington, D.C.: U.S. Government Printing Office, 1986), p. 83.

46. See, for example, Robert Gillette, "Soviets to Give Safety Data on Cuban Reactors to the U.S.," *Los Angeles Times*, Sept. 26, 1986, p. 20; "Soviets Agree to Provide Data on 2 Cuban Reactors," *Washington Post*, Sept. 26, 1986, p. A32.

47. *Granma*, Sept. 30, 1974, p. 2.

48. *Granma*, Nov. 17, 1973, pp. 2-6.

49. "Cumplirá la URSS compromisos petroleros con los paises socialistas," *Granma*, Dec. 21, 1973, p. 8.

50. Jorge F. Pérez-López, "Sugar and Petroleum in Cuban-Soviet Terms of Trade," in Cole Blasier and Carmelo Mesa-Lago, eds., *Cuba in the World* (Pittsburgh: Univ. of Pittsburgh Press, 1979), p. 282.

51. Import unit value calculated from data in Comité Estatal de Estadísticas, *Anuario Estadístico de Cuba 1985*, p. 413; converted to U.S. dollars at the official peso/dollar exchange rate.

52. Based on official Cuban import data and on estimates of Soviet exports to the world and to CMEA in Pérez-López, "Cuban Oil Reexports," p. 5.

53. Marshall Goldman, "The Soviet Union," in Raymond Vernon, ed., *The Energy Crisis* (New York: Norton, 1976), p. 135.

54. *Ibid*, pp. 132-33.

55. Central Intelligence Agency, *Prospects for Soviet Oil Production*, ER77-10270 (April 1977), and *Prospects for Soviet Oil Production: A Supplemental Analysis*, ER77-10425 (July 1977).

56. On this point see, Thane Gustafson, "Soviet Energy Policy," in U.S. Congress, Joint Economic Committee, *The Soviet Economy in the 1980s: Problems and Prospects*, Part I (Washington, D.C.: U.S. Government Printing Office, 1982), p. 432.

57. Jeremy Russell, "Energy in the Soviet Union: Problems for COMECON?" *World Economy*, September 1981, pp. 305-6.

58. Isabel Gorst, "Soviet Union: Serious Drop in Oil Exports," *Petroleum Economist*, vol. 53 (February 1986), p. 2.

59. See Pérez-López, "Cuban Oil Reexports." p. 12.

60. *Lineamientos económicos y sociales*," p. 29.

61. Banco Nacional de Cuba, *Economic Report,* February 1985, p. 7.

62. Banco Nacional de Cuba, *Economic Report,* March 1986, p. 5.

63. Fidel Castro Díaz-Balart, "CMEA Role in Promoting Science, Technology in Cuba," *Ekonomicheskoye sotrudnichestro stran-chienor SEV,* no. 5 (1980), translated from Russian in Foreign Broadcast Information Service, *Worldwide Report: Nuclear Development and Proliferation,* May 11, 1981.

64. *Lineamientos económicos y sociales,* p. 76.

65. For the assumptions underlying those projections see Pérez-López, "Nuclear Power in Cuba after Chernobyl," p. 86.

9

South Africa

MARGARET F. COURTRIGHT

The nexus between South Africa's energy and security cannot be separated from its position as a pariah in international politics. Notwithstanding efforts to ingratiate itself with the United States and Europe in the early post–World War II period through economic ties and acceptance of military installations such as Britain's Simonstown Base, South Africa was never integrated into the Western alliance despite its strategic location at the tip of continent.[1] As years passed, its international political standing deteriorated. This in turn magnified its vulnerabilities to the vagaries of the international energy market, as some exporters of fuels refused to maintain diplomatic ties. Adjustment proved to be a major challenge, but the white regime has exhibited remarkable energy resilience. As an international outcast, South Africa was forced to think through energy vulnerabilities well before the oil crises of the 1970s forced others to to do likewise. As a consequence, it attained self-sufficiency in some energy sectors.

The challenge emerged early in the postwar period. Even before the Afrikaners' rise to power through the National Party in 1948, South Africa's estrangement from the world manifested itself as early as the first session of the United Nations General Assembly, when India and the Soviet Union attacked Pretoria for mistreatment of its Indian population and later for apartheid. As decolonization swept the African continent, efforts to isolate the white regime accelerated. The newly liberated states severed diplomatic relations to bring to bear international pressure to force modification of racial policies.[2]

Over time, countries both within and outside Africa advocated economic and transportation boycotts to isolate the regime. In 1960, the Conference of Independent African States called for closure of ports and airports to South African transport. In 1963, the U.N. General Assembly implemented an arms embargo, which it made mandatory in 1977. By the mid-1970s, the Program of Action Against Apartheid, supported by some 100 members of the General Assembly, called "for the breaking of all diplomatic relations with South Africa, an arms and oil embargo, the suspension of all nuclear cooperation, the cancelling of all loans, investments and technical assistance, the refusal of landing rights to South African aircraft and the closing of all ports to South African ships."[3]

THE ENERGY CHALLENGE

If boycotts and embargoes were to be imposed, if hostility was to be the general rule, South Africa concluded that dependence on other countries must be minimized. As a consequence, in 1946 it incorporated SASOL, the South African Coal, Oil, and Gas Corporation, to safeguard the country against international oil boycotts. In 1949, it founded the Atomic Energy Board to explore the utility of the atom. Benefiting the country in these efforts were gifts of nature—large reserves of coal and uranium.

Coal is South Africa's primary energy resource. It was first mined in 1864 at Moteno, after diamonds were discovered there, and large-scale exploitation began at the turn of the century, when the country produced close to 1 million tons annually. Today South Africa is one of the world's major coal producers, with reserves approaching 110 billion metric tons, half of which is readily exploitable given current technology. The republic also benefits from the relative ease of mining. While three-quarters of coal mines are underground, one-half of all finds are at relatively shallow depths, less than 100 meters. In the 1980s, coal has supplied roughly 80 percent of the the country's energy needs. With much of the new electrical generating capacity built at the mine shaft, coal fires 96 percent of the 21,000 Mwe (megawatts electric) of the country generating capacity.[4]

Beyond electricity, coal produces liquid and gas fuels. Indeed, South Africa has become a world leader in such technology. SASOL I began production in 1955 utilizing high-pressure Lurgi gasifiers to manufacture a synthetic gas—a mixture of hydrogen and carbon monoxide. After sulphur compounds are removed, the remaining gas becomes the raw material for the Fischer-Tropsch synthesis. Plans for SASOL II emerged during the 1973 oil embargo and for SASOL III after the 1979 fall of the Shah of Iran.[5] The plants went into operation in 1980 and 1982, respectively. Collectively producing 45,000 barrels per day[6] of oil equivalent, these sources today provide 6-8 percent of South Africa's total energy requirements.

South Africa's large uranium deposits are an additional means of safeguarding its energy independence while influencing world energy markets. Estimated to be 191,000 metric tons of natural uranium at $30 per pound and 356,000 MT at $50 per pound, South Africa's reserves total 14% of those in the non-Communist world.[7] As part of the ore that was mined for gold, uranium became part of the country's exploited mineral landscape after the early 1920s. However, the mineral did not become commercial until 1952, when the West Rand Consolidated Mines opened a division to extract uranium as a byproduct of gold mining. Production developed rapidly during the 1950s and then proceeded to roller coaster down, reversing its direction only in the 1970s as nuclear energy production expanded worldwide. From 2,786 tons of U_3O_8 in 1976, South African production rose to 6,131 tons in 1981. Since then, production has declined slightly, reflecting the decline in nuclear plant construction worldwide.

Given its large uranium endowment, coupled with a sophisticated engineering and scientific establishment, it was inevitable that Pretoria would consider the generation of atomic power. The West's energy and weapons requirements allowed South Africa leverage to achieve nuclear agreements for technical cooperation with Britain and the United States.[8] The 1957 accord reached with Washington provided for training of scientists and technicians and the sale of a small research reactor, Safari-I, fueled with American enriched uranium.

The legitimacy of the rationale for nuclear energy in such an energy-rich corner of the world rests on a combination of factors. Although nuclear power plants could not address Pretoria's greatest vulnerability—a cutoff of oil, which supplies the transportation sector principally—energy projections in the 1970s suggested that the atom was more economical over the long term than coal, which could be exported more profitably. Reactors would also serve as an avenue for contact with the West at a time when Western domestic and international orders for nuclear power plants were beginning to decline.

Because atomic power was not labor-intensive—having no need for a large work force to provide a continuous feedstock as in coal-fired plants— nuclear energy diversified Pretoria's dependencies on its black workers. Because nuclear fuels are so efficient, the country would be less vulnerable to labor unrest, notwithstanding precautions Pretoria took to address cutoffs through its several-year stockpile of mined coal. Finally, there was the incentive to emulate the path taken by other advanced industrialized countries, which saw the atom as the the road to technological advancement.

With these incentives, nuclear power plant construction began in 1976 for two 922-Mwe reactors at Koeberg, ordered from the French consortium of Framatome, Alsthom, and Spie Batignolles. Difficulties in obtaining a nuclear core delayed startup. The United States begged off from an early commitment, reflecting a turn in South Africa's fortunes in the Congress. This de facto embargo, which began in 1975, formally terminated in 1980 with the application of the Nuclear Non-Proliferation Act. Failing in its effort to manufacture a core, Pretoria persuaded France to deliver one under threat of nonpayment for the reactors.[9] Today the plants supply roughly 4 percent of Pretoria's energy requirements.

Beyond the atom and coal, indigenous petroleum, hydroelectric fuel-wood, and petroleum substitutes account for minor contributions to South Africa's energy profile. Solar electrical generation is negligible. Petroleum remains the republic's Achilles heel. Prospectors have explored the country intensively for oil, to little avail. Offshore drilling has been more successful, but the finds, such as the borehole at Mossel Bay, producing 900 barrels of oil and 90,600 cubic meters of gas per day, have been commercially insignificant.

South Africa thus remains dependent on petroleum imports. To meet the challenge of embargoes, it has stockpiled several years' supply of oil. To address long-term needs, Pretoria has pursued petroleum substitutes. The Department of Agricultural Services has promoted sunflower oil as a diesel fuel substitute in rural areas, but it has undertaken little actual production for this purpose. Ethanol, obtained from fermentation of vegetable matter, including sugar cane, wasted wood, and maize, has received more attention. Sentrachem, a major chemical firm, plans to build ten ethanol plants to produce 1 million tons per year. This could serve as a substitute for 10 percent of the republic's diesel and petroleum needs in the future.

The three hydroelectric stations contribute less than 1 percent of the country's total energy requirements. The Henrik Verwoerd plant (320 Mwe and Vanderkloof installation (220 Mwe) along the Orange River and the Drakensberg station (1000 Mwe) along the Tugela River are uneconomic for the production of electricity alone. However, they make sense as a means of water storage and as a supplement to electricity generation at times of peak load. Fuelwood is a traditional source of energy in many rural areas, with utilization rising slightly through the 1970s and 1980s to about 2.7 million metric tons of coal equivalent. Such figures probably understate actual use, since some wood is simply gathered as needed, a practice that has led the government to complain that savannah forests are being depleted.

SOUTH AFRICA AND THE INTERNATIONAL ENERGY MARKET

Because South Africa is both importer and exporter of energy, it is less vulnerable to international price gyrations than other states in this book. Still, its reliance on imported petroleum subjects it to political gyrations. Pretoria does not publish import statistics; neither do exporters volunteer. Petroleum supplies roughly 12 to 15 percent of the country's energy needs. Until the energy crises of the 1970s, the republic's vulnerability was not particularly acute because of the failure of earlier embargos, proposed by the Conference of Independent African States in June 1960, by the Organ-

ization of African Unity (OAU) at its inception in 1963, and by the United
Nations, beginning with a resolution 1899 in November 1963.

With the 1973 oil crisis, the challenge of obtaining stocks from abroad
became more acute. In November 1973, the OAU, meeting in Addis
Ababa, proposed a five-point program linking African support for the Arab
cause against Israel with the struggle against minority rule in southern
Africa. The OAU secretary-general pointed out that 9 percent of South
Africa's oil imports came from the Persian Gulf and declared that the time
had come for Arab states to use the oil weapon against the white regimes of
southern Africa. The following week, an Arab summit conference at Algiers,
angered by South African support for Israel during the war, adopted a
resolution that included a call for a total Arab oil embargo against Pretoria.

In spite of these calls, South Africa maintained access to Iranian oil,
albeit at higher prices, until the 1979 revolution. Prior to the Shah's fall,
Tehran provided 85 percent to 90 percent of South Africa's petroleum
imports.[10] This ended with the Khomeini regime. Further complicating
matters were "end-user" clauses in Arab contracts with transnational
companies, forbidding the shipment of oil to Pretoria. However, Arab
states—notably Saudi Arabia, Iran, and Oman—never fully implemented
the embargo.

Pretoria addressed the crisis by relying on reserves backed up by
conservation measures, price increases, rationing, and investments in
petroleum substitutes. As time passed and oil shortfalls became oversupply in world markets, Pretoria found it even less difficult to gain access by
paying premium prices and by relying on transshipments. Still, the matter
of access remains a continuing challenge, all the more so since consumption is increasing, particularly among urban blacks.

Oil is not the only energy dependency Pretoria has confronted. Notwithstanding its domestic uranium reserves, it has had to rely on world
markets for high-and low-grade enriched uranium to fuel its research and
power plants, respectively. The United States has terminated its exports in
an effort to thwart the regime's nuclear weapons ambitions and to protest
apartheid. South Africa has turned instead to European suppliers, notably
Belgium and Switzerland, perhaps China as well. Meanwhile, it is increasing domestic production capacity.

South Africa has also chosen to rely on hydel power from neighboring
Mozambique's Cabora Bassa hydroelectric project on the Zambezi River.
Construction began while Mozambique was still a Portuguese colony.
South Africa provided substantial financial support, which continued
following Mozambiquan independence. The productive capacity of the
project is rated at 4,000 Mwe, but so far it generates considerably less. In
1979, one of the last "normal" years, the Electricity Supply Commission

bought 1,400 Mwe from Cabora Bassa, representing 6 percent of South Africa's electrical requirements. During the following decade, there were brownouts in the Transvaal, resulting from attacks on transmission lines by the Mozambique National Resistance and power interruptions by the Mozambique government to protest Pretoria's support for antigovernment guerillas. An improvement in relations has since allowed supply to re-sume—to the benefit of both nations since plant output exceeds Mozam-bique's domestic needs. But because of Mozambique's role as a Front Line State and its past willingness to accept financial losses in its opposition to apartheid, this source of energy is likely to remain unreliable.

As an energy-rich country, South Africa exercises leverage in interna-tional energy markets, with coal being a major export. Transactions in coal steadily increased during the 1970s energy crises, reaching 29.1 million tons per year in the mid-1980s. The growth in exports reflected rising world prices for both oil and coal, the latter increasing from R6 per ton in 1973 to R 20.9 per ton by 1978.[11] By the end of the 1970s, the European Community was importing approximately 23% of its coal from South Africa, reflecting the republic's one-quarter contribution to the global coal export market. When political pressures cut into the market in the 1980s, South Africa compensated by shifting exports to former U.S. markets in the Far East, although at the expense of extensive price cuts.

Pretoria's large uranium reserves have made it a major player on the international uranium market also. Although the country does not publish uranium export data, some estimate that South Africa supplies approx-imately half the uranium oxide needs of European nuclear power plants as well as an important fraction of Japan's requirements. During the 1960s, when uranium oxide prices were depressed, South Africa decided that enriched uranium would be more profitable export than uranium ore. In 1970, it announced a new enrichment procedure—a modification of the German jet-nozzle process—and the formation of the Uranium Enrich-ment Corporation to develop it.[12] In 1975, a small pilot enrichment plant at Valindaba began operations with the capability of producing 40,000 separative work units per year, although it does not appear to have oper-ated at this level. In February 1978, Pretoria laid plans for a "semicommer-cial" plant following a halt in the U.S. supply of enriched fuel for Safari I. Built near Valindaba with the capacity to produce 50 metric tons a year, the plant went into operation in 1988, helping to fuel the Koeberg reactors.

THE NUCLEAR ENERGY—NUCLEAR WEAPONS LINK

South Africa's uranium reserves and manifest technological expertise in enrichment technology supply it with the feedstock for a potential nuclear weapons program. Given a scientific community expert in explosives and the availability of elementary nuclear weapons design in the world's public literature, there is little doubt that Pretoria has the ability to manufacture nuclear weapons if it chooses. It is an open question whether is has already assembled a full-blown nuclear device or the component parts of one.

The Valindaba enrichment plant has been off-limits to international inspection, on the grounds that it would compromise South Africa's enrichment technology. Suspicion focuses on its possible role in a weapons program. Pretoria gave credence to these concerns in 1972 and 1973 annual reports of the Atomic Energy Board. The 1972 report declared that investigations into the "peaceful application of nuclear explosions were being pursued." Although no such research was cited in pronouncements after 1973, the absence has not allayed concerns about the republic's intentions.

The matter of South Africa's nuclear weapons ambitions has been of particular international concern since the late 1970s. Three events fueled speculations. The first was an August 1977 Soviet satellite observation, later confirmed by the United States, that engineering work in the Kalahari Desert bore an uncommon resemblance to preparations for an underground nuclear detonation. When the United States, France, Great Britain, and West Germany demanded explanations, Prime Minister Vorster denied that there was a test site and claimed that "South Africa did not have, nor did it intend to develop, a nuclear explosive device for any purpose, peaceful or otherwise."[13]

The second event was the September 22, 1979, observation by the American Vela satellite of a double flash of light characteristic of a low-yield nuclear explosion. The observation took place in the South Atlantic, off the South African coast, but was not confirmed by other evidence such as radioactive debris. Officially, it remains a mystery. The third occasion was Pretoria's April 1981 anouncement that it had produced uranium enriched to 45%, which would fuel the Safari I research reactor. The announcement raised concern because improving upon this enrichment to a weapons-grade 93 percent does not impose insuperable technological barriers, and American analysts conclude that Pretoria's ability to produce nuclear weapons may be dated to 1980-81.[14]

Other indices of Pretoria's intentions may be found in its refusal to sign the Nuclear Non-Proliferation Treaty (NPT). Pretoria had expressed support for the NPT draft resolution, declaring that "as one of the major producers of uranium in the Western world, South Africa would do absolutely nothing in the context of uranium sales to foreign buyers which might conceivably contribute to an addition to the ranks of the nuclear weapons states." In recent years, to maintain its membership in the International Atomic Energy Agency, South Africa has stated its readiness to join the treaty, but as of this writing has yet to do so. Its failure to sign the accord, coupled with its racial policies, contributed to its ouster from its permanent position on the AEA board of governors in June 1977, and it has not participated in the organization's general conference since 1979, when its credentials were rejected.

Several motives lay behind Pretoria's antagonism toward the NPT. Beyond its official pronouncements that the accord was discriminatory and would exclude achievement of the peaceful benefits of the atom, South Africa, as a uranium producer, was also concerned that an open-ended commitment to allow international safeguards would infringe on its economy. Defense, however, was the principal unstated concern. Pretoria had grounds to doubt that the international community would come to its assistance in the event it was threatened by a nuclear weapons state, notably the Soviet Union, an active supporter of liberation movements.

Assuming that South Africa has available to it highly enriched weapons-grade uranium from its Valindaba plant—which, unlike the Safari I research reactor, is not subject to safeguards—what are the incentives and disincentives to going forward with a military nuclear program? Some see simply capability as incentive enough. Leverage in international diplomacy is another. Even the threat to go nuclear, if not in fact doing so, may provide Pretoria with the club needed to extract concessions from the West.

The military utility of the weapon is more dubious. To be sure, if South Africa were challenged by the Soviet Union or some other military force, the bomb might be a useful deterrent.[15] But with the withdrawal of Cuban forces from Angola, such challenges seem remote. Against guerilla operations—the immediate threat—nuclear weapons would have little utility. The radiological consequences of use on South African territory weighs on the side of caution. As for use against neighboring states, this would bring precisely the kind of reaction South Africa has long tried to prevent: universal ostracism and a uniting of countries, including the superpowers, against the regime. Further, the nuclear capability seems unnecessary, given South Africa's sophisticated conventional armed forces and a military industry that has produced the Cheetah fighter aircraft.

Pretoria might chose to go forward with a nuclear test, however, under the stimulus of political and bureaucratic pressures. Elevating the morale of the white population through a detonation is a conceivable motive. To mitigate international reaction, Pretoria might rely on India's explanation of the weapon as a "peaceful nuclear explosion." It is questionable, however, whether the world reaction would be as muted as it largely was in the Indian case, because Pretoria's position, both politically and strategically, bears little semblance to that of New Delhi. Still, the white regime may yet see the benefits as worth the cost, concluding that this is the only path to the world's respect.

Could outside pressure dissuade Pretoria? It may have worked in 1977, when the Washington intervened to prevent what it believed were preparations for a nuclear detonation. Whether a similar exercise would be successful now is more questionable, since Washington's political leverage has diminished as it has reduced ties. Could neighboring states exercise leverage unavailable to the West? It is unlikely. With economic ties including laborers from Malawi, Mozambique, and Lesotho working in South Africa, with the importation of petroleum products from South Africa, and with South Africa ports serving at times as conduits for their exports, neighboring countries already have more economic involvement with Pretoria than they prefer. Another disincentive might be South African concern about nuclear plants as targets of terrorists. Indeed, such incidents have been reported in Koeberg, although none resulted in serious damage to the plants.[16] The similar risk that a radical white faction within the country could gain access to a weapon or fissile material[17] is a problem that the government believes it can protect against.

Because of its abundant natural resources and its sophisticated scientific establishment, South Africa is more fortunate than many industrializing states in its ability to fashion its energy future. To the extent that it is dependent, particularly on oil to meet its transportation needs, it has built up reserves to meet emergencies. Because it makes major contributions to the world market in coal and uranium, it can exercise counter-leverage.

In the nuclear realm, efforts to isolate Pretoria from the international market at best proved successful only in the very short run. Cut off from Western assistance, Pretoria used its technical resources to acquire a degree of nuclear independence through a sophisticated enrichment capability—a window of opportunity to develop nuclear weapons. Whether it chooses to open this window remains to be seen.

NOTES

1. However, there appears to be fairly close cooperation between the United States and other NATO countries and South Africa, arising out of the Silvermine facility which channels communications obtained from tracking ships around the cape.

2. For example, Zaire broke relations in July 1960, Kenya in 1963, and Zambia in 1967. Cooperation in regional organizations such as the Commission for Technical Cooperation also ended.

3. F. Clifford-Vaughn, ed., *International Pressures and Political Change in South Africa* (Cape Town: Oxford University Press, 1978), p. 5.

4. Republic of South Africa, *Yearbook, 1984*, pp. 445, 578.

5. *London Times,* July 6, 1982.

6. Martin Bailey and Bernard Rivers, *Oil Sanctions Against South Africa* (New York: United Nations Center Against Apartheid, 1978). See also the *Financial Mail,* Dec. 10, 1976, quoting the managing director designate of SASOL, Johannes Stegmann, and the *Rand Daily Mail,* April 14, 1976, quoting the assistant general manager of SASOL. See also Robert Rothberg, *Suffer the Future* (Cambridge: Harvard University Press, 1980), and the *London Times,* July 6, 1982.

7. United Nations, *United Nations Statistical Yearbook,* 1979/80, p. 207.; Mineral Bureau, Republic of South Africa, "South Africa, Persian Gulf of Miners," South Africa Embassy, Washington D.C., July 1982.

8. Richard E. Bissell, *South Africa and the United States: The Erosion of an Influence Relationship* (New York: Praeger, 1982).

9. Dialogue Avec Jean-Pierre Lot, Université et Developement Solidaire (Paris: Berger-Leberalault et Institute International E'tudes Sociales, 1982), p. 31; "Pointers," in *Africa Confidential,* Jan. 6, 1982.

10. *Daily Telegraph,* Feb. 24, 1979.

11. Rothberg, *Suffer the Future,* p. 118.

12. *Yearbook, 1984,* p. 729.

13. In response to a question on ABC network television, "Issues and Answers."

14. Leonard S. Spector, *The Undeclared Bomb* (Cambridge, Mass.: Ballinger, 1988), p. 288.

15. Richard K. Betts, "A Diplomatic Bomb for South Africa," *International Security,* vol. 2 (Fall 1979), p. 107.

16. On Dec. 20, 1982, the Koeberg plant was subject to four small bombs, resulting in little damage. See Konrad Kellen, "The Potential for Nuclear Terrorism; A Discussion," in Paul Leventhal and Yonah Alexander, *Preventing Nuclear Terrorism* (Lexington, Mass.: Lexington Books, 1986), p. 104.

17. Spector, *Undeclared Bomb,* p. 287.

10

Energy and Security
in Industrializing Nations:
Prospects for the Future

BENNETT RAMBERG

In the aftermath of the oil shocks of the 1970s, the world nexus among energy, economy, and security is indisputable. The oil crisis elicited a stream of scholarship addressing the implications, with attention focused on the United States and its OECD allies.[1] To the extent that Lesser Developed Countries and newly industrializing nations were examined, they usually appeared in aggregate data rather than as case studies.[2] The preceding chapters have sought to fill the void by reviewing at length a unique conglomeration of such nations: India, Pakistan, South Korea, Taiwan, Argentina, Brazil, Cuba, and South Africa.

At first glance, these nations may be indistinguishable from a host of other countries that sought to remain upright despite the energy turbulence of the 1970s. Like others, our group turned to the atom to generate electricity. At the same time, the eight nations here—Cuba excepted—manifested a hidden agenda through the years, in which nuclear power generation was a vital energy resource, to be sure, but also a key to nuclear weapons.[3] Some analysts contend that cases such as ours could use nuclear energy as a pretext for securing weapons or weapons technology and feedstock.[4] Although an arguable point, it is not insignificant that of the nations examined here, only South Korea and Taiwan embrace the Nuclear Non-Proliferation Treaty.[5]

Atomic power does, of course, serve legitimate energy security needs. Over a decade ago, Mason Willrich defined such security as assurance of adequate energy supplies to maintain an economy in a politically acceptable manner.[6] Willrich suggested three paths to energy security. First, stockpiling and rationing are remedies for short-term supply interruptions. Whether stockpiling makes sense depends upon the economic significance of a fuel, the degree of reliance on imports, the diversity and reliability of supply, the availability of foreign exchange, the benefits compared with those of other public works, and—in the case of products such as oil, but not nuclear fuels—the availability of warehousing.

Rationing, an allocation device outside the market system, takes many forms. In the Organization for Economic Cooperation and Development during the 1970s, one method combined gas taxes, thermostat regulation, fuel switching (such as substituting coal or wood for oil), car pooling, driverless Sundays, gas station closures, fuel distribution on the basis of license plate numeration, and speed limits. Although such efforts at sharing the burden provided a veneer of equality, elimination of shortfalls in the end required a resumption of the market mechanism, backed up by foreign supply.

Stockpiling and rationing, however, are Band-Aids. They cannot address a nation's energy needs over the long haul. Security of foreign supply through diversification and interdependence offers a second set of alternatives: the options of a buyer's market, when multiple international energy producers have excess capacity. Willrich suggests that to further encourage such markets, either long-term investment in the importer by the exporter and/or industrial assistance by the importer to the exporter will create mutual dependencies. The importer may manipulate the dependencies by making the abusive energy exporter hostage to its own indiscretions. An importer that responds with investment freezes or nationalization challenges the exporter to inflict a self-induced wound upon its international portfolio.

But such interdependence and other forms of diversification only provide the importer with a means to spread the energy risk. Self-sufficiency, the third alternative, is the sole method that can eliminate risks. Except for a very few well-endowed nations, however, autarchy is a chimera. What domestic resource exploitation, conservation, tariffs, and quotas on imported energy can achieve is a reduction of dependencies, not their elimination.[7]

In the energy security challenges of the eight nations we discuss, per capita energy consumption since 1950 often has come close to doubling every fifteen to twenty years (table 1). To demonstrate what consumption levels might be were our sample states to achieve the "advanced" stage of economic development in industrialized states, table 10.1 includes comparative figures for the United States and France.

Whether per capita consumption will accelerate for countries strapped with large domestic debts and balance-of-payments difficulties is uncertain over the short term. Indeed, growth in Argentina, Brazil and Cuba actually declined between 1980 and 1985. All the same, the long-term outlook suggests a steady upward trend because of rising populations and heretofore unmet economic needs.

Table 10.1.   Per capita energy consumption
               (Kilograms of oil equivalent)

	Argentina	Brazil	Cuba	India	Pakistan	S. Korea	S. Africa	Taiwan[a]	France	U.S.
1950	502	124	295	30	—	30	924	—	1,282	5,140
1960	727	210	593	74	—	134	1,241	409	1,558	5,535
1970	1,087	299	686	95	—	450	1,510	702	2,639	7,436
1975	1,138	453	867	111	112	625	1,680	1,006	2,597	7,199
1980	1,202	526	982	126	136	941	1,892	1,622	3,010	7,163
1985	1,192	484	748	178	175	1,130	2,147	1,528[b]	2,909	6,694

Sources: United Nations, *1982 Energy Statistics Yearbook*, New York, 1984, pp. 61, 68, 71, 73, 79, 81, 83; ibid, *1985 Energy Statistic Yearbook*, New York, 1987, pp. 37, 39, 43, 47, 49, 53; *Statistical Year Book of the Republic of China*, Director-General of Budget, Accounting and Statistics, 1984, p. 324.
[a]Per capita consumption per liter.
[b]1982 per capita consumption.

The impact on world energy markets is evident in figures for the period between 1970 and 1985, when the Less Developed Countries' fraction of world energy consumption (excluding traditional biomass fuels) increased from 14 percent to 23 percent. For the same period, the LDC share of world oil supplies rose from 13 percent to 23 percent.[8] Changing life styles are major stimuli for such growth. Urbanization inspires energy-intensive commercial, industrial, and transportation services. It particularly expands residential electrical demands for lighting, appliances, refrigeration, and air conditioning. LDC demand for natural gas for water and space heating and cooking has increased generally by 10-20 percent annually. In Pakistan, the growth was 25 percent per annum during 1972-84.[9]

Such growth is not surprising when one considers that in rural Brazil, India, and Pakistan, for example, dependencies on biomass such as firewood run between 90 and 95 percent.[10] With urbanization, these needs must be met from other sources. In Taipei's rapidly expanding economy, for one, the number of television sets increased from almost nil to seventy-seven per hundred households between 1962 and 1972; rice cookers from five to ninety-nine; and refrigerators from two to forty-eight. Although growth slowed in the following decade, in large measure this reflected saturation of the domestic market. For example, by 1982 refrigerators could be found in 92 percent of Taiwanese urban households. By contrast, other residential amenities, notably energy-intensive air conditioning, could be found in only 23 percent of city dwellings in 1982, allowing for a significant expansion in energy demand to meet this need in the years ahead.[11] Other energy demands are also foreseeable, one being the

Table 10.2.   Imports as a percentage of total energy utilization

	Argentina	Brazil	Cuba[a]	India	Pakistan	S. Korea[b]	S. Africa	Taiwan	France	U.S.
1950	71.9	78.5	100.6	12.2	—	50.0	13.9	—	43.5	5.5
1960	40.3	68.9	104.7	22.5	—	20.4	14.8	16.1	61.4	9.5
1970	11.9	67.3	102.2	23.5	—	62.5	29.1	50.0	90.4	12.4
1975	17.1	77.6	105.3	23.0	51.3	76.6	32.9	66.5	100.0	20.8
1980	15.0	76.3	106.0	25.9	46.1	88.0	27.1	83.6	102.1	22.5
1985	6.4	51.4	122.0	14.9	40.8	90.5	21.9	83.2	85.6	16.1

Sources: United Nations, *1982 Energy Statistics Yearbook*, New York, 1984, pp. 60-61, 64-65, 70-71, 62-73, 78-79, 80-81, 82-83, 86-87; ibid., *Energy Statistics Yearbook*, 1987, pp. 24-25, 36-37, 38-39, 42-43, 46-47, 48-49; *1985 Statistical Year Book of the Republic of China*, Director-General of Budget, Accounting and Statistics, p. 324.

[a]The 100%-plus figures for Cuba reflects utilization of oil imports for energy as well as other industrial purposes.

[b]Official Korean statistics place reliance on imported energy in 1986 at 66.5% to rise to 77.6% by 2010. See Ton Wan Park's article in this book.

growing need for gasoline to fuel an expanding use of the automobile in Asian countries that today rely on motorbikes.

When energy dependencies in our case nations are stated in terms of imports as a fraction of total consumption, two Latin American nations, Argentina and Cuba, define the boundaries (table 10.2). Cuba imports virtually all its energy supplies, as it has done for three decades. By contrast, Argentina made remarkable progress toward self-sufficiency during the same period. During 1960-82, Buenos Aires reduced foreign dependence from 64 percent to only 8 percent. Brazil and South Africa made modest progress, while the remainder of our sample cases have either experienced no important gains or, in the case of South Korea and Taiwan, increased substantially their dependence on imports.

In one fashion or another, the nations examined here have applied Willrich's three sets of alternatives. To cushion the impact of supply interruptions, there are at least two instances of substantial energy stock-piling: South Africa maintains a three-year storage of petroleum and South Korea, a 150-day supply. There is a spectrum of efforts aimed at diminishing energy dependencies through diversification. Cuba lies to one extreme, deriving 99 percent of its needs from the Soviet bloc; South Korea, on the other end, expanded its sources from seven oil exporters in 1980 to thirteen in 1982. The remaining case nations range across this spectrum.

In diversification or changes in the composition of energy profiles, nuclear power is prominent. Imported nuclear energy plants, which played either a minor or nonexistent role a decade ago, are coming into their own (table 10.3). With fifteen plants, India has the largest number in

Table 10.3.   Nuclear power plants at end of 1988

	Operating		Under construction		Planned		Operating units as percentage of total capacity
	units	Mw(e)	units	Mw(e)	units	Mw(e)	
Argentina	2	935	1	692	—	—	13.4
Brazil	1	657	2	2,618	6	7470	0.05
Cuba	0	0	2	880	0	0	0
India	7	1,243	8	1,880	3	1,235	2.6
S. Korea	9	5,816	1	920	2	1,900	53.1
Pakistan	1	137	0	0	1	900	1.0
S. Africa	2	1,800	0	0	0	0	4.5
Taiwan	6	5,144	0	0	2	2,000	48.5

Source: *Nuclear News*, February 1989, pp. 69, 70, 74, 76, 82.

operation, under construction, and planned. Korea produces the most nuclear energy and is the most reliant, with 53 percent of its electricity coming from this source. In the years to come, atomic power will continue to grow in many case nations. Indeed, except for Cuba, they are likely to become exporters of nuclear components.[12] India and Argentina, for example, today manufacture much of their hardware and are beginning to produce nuclear fuel.[13]

Apart from nuclear energy, several countries rely on interdependencies to diminish their vulnerabilities to supply interruptions. South Africa has invested heavily in a hydroelectric project in Mozambique upon which it relies and has also attempted to expand a web of European dependencies on its uranium reserves. Korea operates joint energy ventures in coal and oil abroad and uses its arms industry to gain a foothold in Arab states, thereby creating reciprocal dependencies. Brazil exports technology and oil exploration equipment to Iraq, Iran, and Colombia and has provided nuclear assistance to Iraq. A Brazilian oil venture in Iraq resulted in the discovery of the major Majnoon oil field. Demonstrating the fragility of such ventures, however, Iraq in 1980 took full possession of the field; in lieu of the usual production-sharing agreement, Baghdad offered a guaranteed amount of crude deliveries.[14]

Beyond interdependence, many of our case nations follow Willrich's third option, development of untapped domestic resources. This alternative includes such unconventional technologies as coal liquefaction, which provides South Africa with 6 percent of its energy. Among other substantial efforts, Brazil is converting sugar cane to ethanol to provide fuel for the transportation sector.[15]

Table 10.4.   Energy profiles

	Published proven reserves of natural gas, 1984 (million cu m)	Published crude petroleum reserves, 1987 (thousand bbl)	Years of proven oil reserves	Coal reserves (thousand MT)
Argentina	676	2,180	13	130
Brazil	95	2,358	11	2,343
Cuba	—	—	—	—
India	575	4,375	20	1,581
Korea	—	—	—	132
Pakistan	625	116	8	102
So. Africa	8	115	?	58,404
Taiwan	21	10	10	200
France	33	222	11	381
U.S.	5,202	27,280	9	263,843

Source: *Britannica Book of the Year, 1988* (Chicago: Encyclopedia Britannica, 1988), pp. 806-11.

Our cases also exploit traditional domestic resources for conventional purposes, each country approaching this with a different profile (table 10.4). Some, notably South Africa and India, are relatively rich in resources such as coal. In most cases, the reserves of natural gas, petroleum, or coal are modest at best. In Cuba, by contrast, such reserves are absent. South Africa has large coal and uranium deposits, and India, large untapped hydroelectric potential and coal. Both Brazil and Argentina have unexploited hydroelectric potential. Korea has modest coal deposits, as does Taiwan. Pakistan's extensive coal deposits are largely of poor quality, while its hydroelectric potential lies at some distance from population centers. To the extent it can, each case is attempting to develop domestic resources within the limits of available capital and competing public works.

Given domestic capital limitations, many of our cases will require the infusion of foreign investment to develop domestic potential. For this to occur, hostility toward foreign development of natural resources must be overcome. Resistance, however, is dissolving only slowly. Brazil, which opposed such involvement some five decades, has opened its continental shelf to external exploration. At first the financial incentives for investors were so restrictive that there were few bidders, but after a time, Brazil loosened its terms.[16]

As each nation tries to meet its energy needs, several patterns are foreseeable. Atomic power will play a role as a civilian veneer for a nuclear weapons option, while it has the potential to diversify energy profiles and thereby make the case nations more energy-secure. However, atomic power, growing much more slowly than anticipated, cannot make up the energy shortfall, and each nation will pursue a distinct strategy to meet its needs. South Africa, in addition to developing its large domestic coal resources, will seek to exploit Western and neighboring dependencies on its energy and mineral resources. India has important domestic reserves of coal and hydel power. South Korea and Taiwan will expand and diversify their reciprocal dependencies with oil exporters. Pakistan will look to increase efficient use of its oil, natural gas, and hydel potential. Brazil will substantially expand its commitment to ethanol and slowly enlarge its hydel potential, as will Argentina. Cuba's domestic options are limited; even with nuclear generating plants, it will remain heavily reliant on the Soviet bloc. For each case nation, energy security will remain a preoccupation for the foreseeable future.

NOTES

1. Numerous books have addressed the energy crisis, with a clear focus on the OECD experience. A very brief list includes: Guy de Carmoy, *Energy for Europe: Economic and Political Implications* (Washington, D.C.: American Enterprise Institute, 1977); Melvin Conant, *The Oil Factor in U.S. Foreign Policy* (Lexington, Mass.: Lexington Books, 1982); David A. Deese and Joseph S. Nye, eds., *Energy and Security,* (Cambridge, Mass.: Ballinger, 1981); Martin Greenberger, *Caught Unawares: The Energy Decade in Retrospect* (Cambridge, Mass.: Ballinger, 1983); Wilfred L. Kohl, *After the Second Oil Crisis* (Lexington, Mass.: Lexington Books, 1982); Hans Landsberg, ed., *Energy: The Next Twenty Years* (Cambridge, Mass.: Ballinger, 1979); and Robert Staugaugh and Daniel Yergin, eds., *Energy Future: Report of the Energy Project at the Harvard Business School* (New York: Random House, 1979).

2. Typical is chapter 4 in International Energy Agency, *World Energy Outlook* (Paris: Organization for Economic Cooperation and Development, 1982).

3. Albert Wohlstetter, et al., *Moving Toward Life in a Nuclear Armed Crowd,* Los Angeles: Pan Heuristics, 1976.

4. Ibid.

5. Chauncey Starr and E. Zebroski, "Nuclear Power and Weapons Proliferation," paper prepared for the American Power Conference, April 18, 1977.

6. Mason Willrich, ed., *Energy and World Politics* (New York: Free Press, 1975). chap. 4.

7. James W. McKie, "Domestic Aspects of Energy Security Policy," in Richard J. Gonzalez et al., eds., *Improving U.S. Energy Security* (Cambridge, Mass.: Ballinger, 1985). chap. 11.

8. Jayant Sathaye et al., "Energy Demand in Developing Countries: A Sectoral Analysis of Recent Trends," in Jack M. Hollander et al., *Annual Review of Energy,* vol. 12 (Palo Alto, Calif.: Annual Reviews Inc., 1987), pp. 253, 260, 263, 273, 277, 378.

9. Gerald A. Leach, "Residential Energy in the Third World," in Hollander et al., *Annual Review of Energy,* vol. 13, 1988, p. 48. Note that in LDCs 30-90 percent of total energy use is residential, compared with 25-30 percent in industrialized countries. Presumably, figures for industrializing countries in our sample countries fall somewhere in between.

10. Ibid, p. 52.

11. Jayant Sathaye and Stephen Meyers, "Energy Use in Cities of the Developing Countries," in Hollander et al., *Annual Review of Energy*, vol. 10, 1985, p. 123.

12. Rodney W. Jones et al., eds, *The Nuclear Suppliers and Nonproliferation* (Lexington, Mass.: Lexington Books, 1985).

13. See William C. Potter, ed., *International Nuclear Trade and Nonproliferation* (Lexington, Mass.: Lexington Books, 1990).

14. Paul Kemezis and Ernest J. Wilson III, *The Decade of Energy Policy: Policy Analysis in Oil-Importing Countries* (New York: Praeger, 1984), p. 111.

15. Howard S. Geller, "Ethanol Fuel From Sugar Cane in Brazil," in Hollander, *Annual Review of Energy*, vol. 12, pp. 135-64.

16. Kemezis and Wilson, *Decade of Energy Policy*, pp. 140-42.

Contributors

Bennett Ramberg, B.A. (Southern California); M.A., Ph.D. (Johns Hopkins): J.D. (UCLA), is senior research associate at the Center for International and Strategic Affairs, UCLA. He was earlier a research fellow at Stanford University's Arms Control Program and Princeton University's Center for International Studies. He has also served as a foreign policy consultant to government. Dr. Ramberg is the author of *Global Nuclear Energy Risks: The Search for Preventive Medicine* (Westview, 1986); *Nuclear Power Plants as Weapons for the Enemy: An Unrecognized Military Peril* (University of California Press, 1984); *Globalism vs. Realism: International Relations' Third Debate*, coedited with Ray Maghroori (Westview, 1982); *Destruction of Nuclear Energy Facilities in War: The Problem and the Implications* (Lexington Books, 1980); and *The Seabed Arms Control Negotiations: A Study of Multilateral Arms Control Conference Diplomacy* (University of Denver Monograph Series, 1978). He also has published in leading journals.

Raju G.C. Thomas, B.A., M.A. (Bombay); B.Sc. Econ. (London School of Economics); M.A. (Southern California); Ph.D. (UCLA), is professor of political science at Marquette University. He was formerly research fellow at the Harvard University Center for Science and International Affairs, MIT's Center for International Studies, and the UCLA Center for International and Strategic Affairs. Dr. Thomas also has worked with multinational corporations in Europe and India. His publications include *Indian Security Policy* (Princeton University Press, 1986); *The Defense of India: A Budgetary Perspective* (Macmillan, 1978); and, as editor, *The Great Power Triangle and Asian Security* (Lexington Books, 1983). He has published numerous journal articles.

Margaret F. Courtright, B.A. (Radcliffe), Ph.D. (Johns Hopkins), is a certified public accountant and formerly an assistant professor of political science at Marquette University, specializing in international political economy and African politics. Her major area of interest at Radcliffe was international trade and development. At Johns Hopkins, she studied international law, arms control, and the political economy of South Africa.

Charles K. Ebinger, M.A., M.A.L.D., Ph.D. (Fletcher School of Law and Diplomacy), is associated with the Washington, D.C.–based Center for Strategic and International Studies. He was formerly vice president of Conant Associates, a Washington, D.C.–based energy consulting firm, and has been associated with the American Universities Field Staff. He has written and published extensively on international affairs and energy policy. His two most recent works are the

International Politics of Nuclear Energy (Sage, 1978), and *Pakistan: Energy Planning in a Strategic Vortex* (Indiana University Press, 1981).

Tong Whan Park, LL.B. (University National Seoul); M.A., Ph.D. (University of Hawaii), is associate professor of political science at Northwestern University and principal investigator of the international energy project there since 1973. He has published extensively in leading academic journals on international energy politics, on the politics of Asia and the Third World, and on the political economy of newly industrializing countries. His energy publications include *Energy and Security: Korea's Problems and Tasks in the 1980s* (Research Institute for International Affairs, Seoul, 1980) and *Problems of Petroleum Supply to Korea* (Institute for Foreign Affairs and National Security, Seoul, 1979). He is currently conducting an analysis of the arms race, deterrence, and coexistence between the two Koreas in the context of the emerging quadrilateral balance of power in Northeast Asia.

Jorge F. Pérez-López, B.A. (State University of New York, Buffalo); M.A., Ph.D. (State University of New York, Albany), is an international economist with the Bureau of International Labor Affairs, U.S. Department of Labor; he was previously associated with the department's Bureau of Labor Statistics. He has taught at the School of Business, State University of New York at Albany. Dr. Pérez-López has written and published extensively on various topics related to U.S. trade policy, the measurement of OPEC-induced inflation, and various aspects of the Cuban economy. He is contributing editor of the *Handbook of Latin American Studies* for the section on the Cuban economy.

Denis Fred Simon, B.A. (State University of New York, New Paltz); M.A., Ph.D. (U.C. Berkeley), is professor of international relations at the Tufts University Fletcher School of International Law and Diplomacy, where he conducts research dealing with international technology transfer, foreign investment, and East Asian development. He was formerly a research fellow at the East-West Center, University of Hawaii, working on energy and technology problems in the Asian region. He has also taught at George Mason University and the University of Virginia. He is the author of *Taiwan, Technology Transfer and Transnationalism: The Political Management of Dependency* (Westview, 1984) and has published several articles on the problems of technological development in China and Taiwan.

Etel Solingen, B.A., M.S. (Hebrew University, Jerusalem); Ph.D. (UCLA), is assistant professor in the Department of Politics and Society at the University of California, Irvine. She held a University of California Institute on Global Conflict and Cooperation postdoctoral award, was a UCLA International Studies and Overseas Program postdoctoral fellow (1989) and taught at the University of Southern California's School of International Relations (1988). She is also senior research fellow at UCLA's Center for International and Strategic Affairs and Latin American Center. Her publications include comparative studies of political determinants of scientific and technological change in industrializing states, and she has recently completed a manuscript on technology and the industrializing state, focusing on Brazil's nuclear industry.

Cynthia A. Watson, B.A. (University of Missouri-Kansas City), M.A. (London School of Economics), Ph.D. (University of Notre Dame), is assistant professor of political science at Loyola University of Chicago, with a specialization in Latin American politics and security studies. Her major area of research at the London School of Economics was the economic history of Argentina and Colombia. At Notre Dame, she worked on nuclear questions and their application to Argentina and its foreign affairs. She currently is writing a study comparing the military's response to political violence in Argentina and Colombia.

Index

CISA Publications

The CISA Book Series: "Studies in International and Strategic Affairs," includes the following:

William C. Potter, ed., *Verification and SALT* (Westview Press, 1980).

Bennett Ramberg, *Destruction of Nuclear Energy Facilities in War: The Problem and Implications* (Lexington Books, 1980); revised and reissued as *Nuclear Power Plants as Weapons for the Enemy: An Unrecognized Military Peril* (University of California Press, 1984).

Paul Jabber, *Not by War Alone: Security and Arms Control in the Middle East* (University of California Press, 1981).

Roman Kolkowicz and Andrzej Korbonski, eds., *Soldiers, Peasants, and Bureaucrats* (Allen & Unwin, 1982).

William C. Potter, *Nuclear Power and Nonproliferation: An Inter-disciplinary Perspective* (Oelgeschlager, Gunn, and Hain, 1982).

Steven L. Spiegel, ed., *The Middle East and the Western Alliance* (Allen & Unwin, 1982).

Dagobert L. Brito, Michael D. Intriligator, and Adele E. Wick, ed., *Strategies for Managing Nuclear Proliferation—Economic and Political Issues* (Lexington Books, 1983).

Bernard Brodie, Michael D. Intriligator, and Roman Kolkowicz, eds., *National Security and International Stability* (Oelgeschlager, Gunn and Hain, 1983).

Raju G.C. Thomas, ed., *The Great Power Triangle and Asian Security* (Lexington Books, 1983).

R.D. Tschirgi, *The Politics of Indecision: Origins and Implications of American Involvement with the Palestine Problem* (Praeger, 1983).

Giacomo Luciani, ed., *The Mediterranean Region: Economic Inter-dependence and the Future of Society* (Croom Helm and St. Martin's Press, 1984).

Roman Kolkowicz and Neil Joeck, eds., *Arms Control and International Security* (Westview Press, 1984).

Jiri Valenta and William C. Potter, eds., *Soviet Decisionmaking for National Security* (Allen & Unwin, 1984).

William C. Potter, ed., *Verification and Arms Control* (Lexington Books, 1985).

Rodney Jones, Joseph Pilat, Cesare Merlini, and William C. Potter, eds., *The Nuclear Suppliers and Nonproliferation: Dilemmas and Policy Choices* (Lexington Books, 1985).

Gerald Bender, James Coleman, and Richard Sklar, eds., *African Crisis Areas and U.S. Foreign Policy* (University of California Press, 1985).

Bennett Ramberg, *Global Nuclear Energy Risks: The Search for Preventive Medicine* (Westview Press, 1986).

Neil Joeck, ed., *Strategic Consequences of Nuclear Proliferation in South Asia* (Frank Cass, 1986).

Raju G.C. Thomas, *Indian Security Policy* (Princeton University Press, 1986).

Steven L. Spiegel, Mark Heller, and Jacob Goldberg, eds., *Soviet-American Competition in the Middle East* (Lexington Books, 1987).

Roman Kolkowicz, ed., *The Logic of Nuclear Terror* (Allen & Unwin, 1987).

Roman Kolkowicz, ed., *Dilemmas of Nuclear Deterrence* (Frank Cass, 1987).

Michael D. Intriligator and Hans-Adolf Jacobsen, eds., *East-West Conflict: Elite Perceptions and Political Options* (Westview Press, 1988).

Marco Carnovale and William C. Potter, eds., *Continuity and Change in Soviet-East European Relations: Implications for the West,* (Westview Press, 1989).

William C. Potter, ed., *The Emerging Nuclear Suppliers and Nonproliferation* (Lexington Books, 1990).